Intelligent Systems Reference Library

Volume 282

The aim of this series is to publish a Reference Library, including novel advances and developments in all aspects of Intelligent Systems in an easily accessible and well structured form. The series includes reference works, handbooks, compendia, textbooks, well-structured monographs, dictionaries, and encyclopedias. It contains well integrated knowledge and current information in the field of Intelligent Systems. The series covers the theory, applications, and design methods of Intelligent Systems. Virtually all disciplines such as engineering, computer science, avionics, business, e-commerce, environment, healthcare, physics and life science are included. The list of topics spans all the areas of modern intelligent systems such as: Ambient intelligence, Computational intelligence, Social intelligence, Computational neuroscience, Artificial life, Virtual society, Cognitive systems, DNA and immunity-based systems, e-Learning and teaching, Human-centred computing and Machine ethics, Intelligent control, Intelligent data analysis, Knowledge-based paradigms, Knowledge management, Intelligent agents, Intelligent decision making, Intelligent network security, Interactive entertainment, Learning paradigms, Recommender systems, Robotics and Mechatronics including human-machine teaming, Self-organizing and adaptive systems, Soft computing including Neural systems, Fuzzy systems, Evolutionary computing and the Fusion of these paradigms, Perception and Vision, Web intelligence and Multimedia.

Indexed by SCOPUS, DBLP, zbMATH, SCImago.

All books published in the series are submitted for consideration in Web of Science.

Manuel Mora · Jorge Marx Gómez · Fen Wang ·
Hector A. Duran-Limon
Editors

Engineering and Management of Data Science, Analytics, and AI/ML Projects

Foundations, Models, Frameworks, Architectures, Standards, Processes, Practices, Platforms and Tools for Small and Big Data

Springer

Editors
Manuel Mora
Information Systems Department
Autonomous University of Aguascalientes
Aguascalientes, Mexico

Jorge Marx Gómez
Department of Informatics
Carl von Ossietzky University of Oldenburg
Oldenburg, Niedersachsen, Germany

Fen Wang
Information Technology Management
Central Washington University
Ellensburg, WA, USA

Hector A. Duran-Limon
Information Systems Department
University of Guadalajara
Zapopan, Jalisco, Mexico

ISSN 1868-4394 ISSN 1868-4408 (electronic)
Intelligent Systems Reference Library
ISBN 978-3-032-06888-0 ISBN 978-3-032-06889-7 (eBook)
https://doi.org/10.1007/978-3-032-06889-7

This Springer imprint is published by the registered company Springer Nature Switzerland AG
The registered company address is: Gewerbestrasse 11, 6330 Cham, Switzerland

Preface

This book aims to provide a dual perspective on modern research and praxis on Data Science, Analytics, and AI/Machine Learning (DSA-AI/ML) projects with small or big data. Academics, researchers and practitioners interested in the systematic development and implementation of Data Science, Analytics, and AI/Machine Learning projects have available chapters reporting Engineering and Management foundations regarding frameworks, architectures, and standards for DSA-AI/ML projects. Readers can also find chapters on updated reviews on platforms and tools for DSA-AI/ML projects. Finally this book provide relevant insights on achievements, challenges and trends of DSA-AI/ML projects.

This book is co-edited through an open public academic call for chapters reported in digital academic international groups in the disciplines of Data Science, Analytics, AI/ML and Software and Systems Engineering to submit high-quality conceptual and/or empirical research chapters on the following—but not limited to—topics of interest:

- **Foundations on Engineering and Management of Data Science, Analytics, and AI/ML Projects**. Topics: fundamental frameworks, architectures, and standards for DSA-AI/ML projects.
- **Development Platforms and Tools for Data Science, Analytics, and AI/ML Projects**. Topics: technical reviews, comparative studies, didactical descriptions, performance studies on the diverse technological resources required for DSA-AI/ML projects.
- **Achievements, Challenges, Trends and Future Research Directions on Data Science, Analytics, and AI/ML Projects**. Topics: Review of real-life achievements, challenges, current problems and limitations, trends, and future research directions.

The editorial process lasted about 18 months for disseminating the open call for chapters, receiving the initial chapter proposals, receiving the first version of the full chapters, conducting the first single blind-mode review, reporting the initial editorial decisions, receiving the second version full chapters, reporting the final editorial decisions, and receiving the final camera-ready version of the final accepted chapters. During this editorial process, we received about 12 initial chapter proposals, but finally six high-quality chapters were completed and accepted after the two review cycles. The final six accepted chapters are from senior researchers located in USA, Germany, Turkey and Mexico countries.

The first initial three chapters provide research-oriented content on foundations for understanding the technological bases of Data Science, Analytics, and AI/Machine Learning projects addressing relevant topics on Non-Proprietary Domain-Independent Data Science Analytics AI/ML Reference Architectures, Data Visualization Principles and Techniques, and Requirements for using Federated Learning in large-scale systems such as Manufacturing Supply Chains. The final three chapters report useful insights on current and potential applications of Data Science, Analytics, and AI/Machine Learning projects on the topics of Large Multimodal Model-Based Simulation Game for Value Stream Analysis and Design using a Socio-Technical System approach, Data-Driven Clustering Approach for Assessing Service Performance Food Service Industry brands, the potential integration of Quality Management 4.0 with AI and Machine Learning Data Analytics Systems.

Hence, we consider that this research-oriented co-edited book entitled *Engineering and Management of Data Science, Analytics, and AI/ML Projects—Foundations, Models, Frameworks, Architectures, Standards, Processes, Practices, Platforms and Tools for Small and Big Data*—contributes on this relevant current research stream and provides useful insights to advance toward the systematic development and implementation of Data Science, Analytics and AI/ML projects.

Aguascalientes, Mexico — Prof. Manuel Mora
Oldenburg, Germany — Prof. Jorge Marx Gómez
Ellensburg, USA — Prof. Fen Wang
Zapopan, Mexico — Prof. Hector A. Duran-Limon

References

1. Cao, L.: Data science: challenges and directions. Commun. ACM 60(8), 59–68 (2017)
2. Davenport, T., Malone, K.: Deployment as a Critical Business Data Science Discipline. Harvard Data Sci. Rev. 3(1) (2021)
3. Haakman, M., Cruz, L., Huijgens, H., van Deursen, A.: AI lifecycle models need to be revised. Empirical Software Eng. 26(5), 1–29 (2021)
4. Huyen, C.: Designing Machine Learning Systems. O'Reilly Media, Inc. (2022)
5. ISO/IEC: ISO/IEC 5338:2023—Information technology—Artificial intelligence—AI system life cycle processes. ISO—International Organization for Standardization, Geneva (2023)
6. Martinez, I., Viles, E., Olaizola, I. G.: Data science methodologies: Current challenges and future approaches. Big Data Res. 24, (2021)
7. Kelleher, J. D., Tierney, B.: Data Science. MIT Press (2018)
8. Mora, M., Wang, F., Marx-Gomez, J., Duran-Limon, H.: Development Methodologies for Big Data Analytics Systems: Plan-driven, Agile, Hybrid, Lightweight Approaches. Springer (2023)
9. OECD: Scoping the OECD AI Principles. OECD, Paris (2019)

Acknowledgments

We thank our following colleagues for your academic effort spent in the single blind-mode double-cycle review process:

- Arturo Chavoya, University of Guadalajara, Mexico
- Doncho Petkov, Eastern Connecticut State University, USA
- Fen Wang, Central Washington University, USA
- Hector A. Duran-Limon, University of Guadalajara, Mexico
- Jonas Kallisch, OFFIS, Germany
- Jorge Marx Gómez, University of Oldenburg, Germany
- Lutz Engel, Jade University of Applied Sciences, Germany
- Mahesh Raisinghani, Texas Woman's University, USA
- Manuel Mora, Autonomous University of Aguascalientes, Mexico
- Mike Geisthardt, Jade University of Applied Sciences, Germany
- Oswaldo Diaz, INEGI, Mexico
- Paola Y. Reyes-Delgado, Autonomous University of Aguascalientes, Mexico
- Sergio Galvan-Cruz, Autonomous University of Aguascalientes, Mexico
- Serhat Peker, Izmir Bakircay University, Turkey

We also thank Prof. Lakhmi C. Jain and Prof. Janusz Kacprzyk, co-Editors of the Springer Nature—Book Series: *Intelligent Systems Reference Library* Series, for all academic support provided for the realization of this book.

Finally, we express our gratitude to Thomas Ditzinger—Senior Editor at Springer Book Series—and Aishwarya Srinivasan—Production Editor at Springer—for their Editorial Staff for all guidance provided for the publication of this book.

Contents

Editors and Contributors

About the Editors

Manuel Mora is a full-time Professor in the Information Systems Department at the Autonomous University of Aguascalientes (UAA), Mexico. Dr. Mora holds an M.Sc. in Computer Sciences (Artificial Intelligence area, 1989) from Monterrey Tech (ITESM), and an Eng.D. in Engineering (Systems Engineering area, 2003) from the National Autonomous University of Mexico (UNAM). He has published over 90 research papers in international top conferences, research books, and JCR indexed journals such as IEEE-TSMC, European Journal of Operational Research, International Journal of Information Management, Engineering Management, International Journal of Information Technology and Decision Making, Information Technology for Development, International Journal in Software Engineering and Knowledge Engineering, Engineering Management Journal, Computer Standards and Interfaces, Software Quality Journal, Expert Systems, Software and Systems Modeling, Journal of Organizational Computing and Electronic Commerce, The Journal of Supercomputing, and Information Systems and e-Business Management. Dr. Mora is a senior member of ACM (since 2008), an SNI at Level II, and serves in the ERB of several international journals indexed by Emergent Source Citation Index focused on decision-making support systems (DMSS) and IT services systems.

Prof. Dr. Jorge Marx Gómez studied Computer Engineering and Industrial Engineering at the University of Applied Sciences Berlin (Technische Fachhochschule Berlin). He was a lecturer and researcher at the Otto-von-Guericke-Universität Magdeburg (Germany) where he also obtained a Ph.D. degree in Business Information Systems with the work Computer-based Approaches to Forecast Returns of Scrapped Products to Recycling. From 2002 till 2003 he was a visiting professor for Business Informatics at the Technical University of Clausthal (TU Clausthal, Germany). In 2004 he received his habilitation for the work Automated Environmental Reporting through Material Flow Networks at the Otto-von-Guericke-Universität Magdeburg. In 2005 he became a full professor and chair of Business

Information Systems at the Carl von Ossietzky University Oldenburg (Germany). His research interests include Very Large Business Applications, Business Information Systems, Federated ERP-Systems, Business Intelligence, Data Warehousing, Interoperability, and Environmental Management Information Systems.

Fen Wang is a Full Professor in the Information Technology and Administrative Management Department at Central Washington University (CWU). Before joining CWU, Prof. Wang was an Assistant Professor and Director of the Management Information Systems (MIS) program at the Eastern Nazarene College in Massachusetts. Professor Wang holds a B.S. in MIS, an M.S., and a Ph.D. in Information Systems from the University of Maryland Baltimore County. Professor Wang has brought over ten years of professional and research experience in information technology management to her students. Her research focuses on intelligent decision support technologies and E-business strategies. These efforts have resulted in contributions to the applied literature on information technologies that have been well-received in the research community. Professor Wang has published over 30 papers in internationally-circulated journals and book series, including the International Journal of E-Business Research (IJEBR), International Journal of Decision Support System Technology (IJDSST), Intelligent Decision Technologies (IDT), Information Technology for Development (ITFD), and the Encyclopedia of E-Commerce, E-Government and Mobile Commerce. Professor Wang has also consulted for a variety of public and private organizations on IT management and applications.

Hector A. Duran-Limon Ph.D., is currently a full Professor at the Information Systems Department, University of Guadalajara, Mexico. He completed a Ph.D. at Lancaster University, England in 2002. Following this, he was a post-doctoral researcher until December 2003. He obtained an IBM Faculty award in 2008. His research interests include Cloud Computing and High-Performance Computing (HPC). He is also interested in Software Architecture, Software Product Lines, and Component-based Development. In 2006, He was invited to create a Ph.D. program in Information Technologies for the University of Guadalajara, becoming a member of the Academic Council. Contact him at the Information Systems Department, University of Guadalajara, Mexico; hduran@cucea.udg.mx.

Contributors

Arturo Chavoya Information Systems, CUCEA, University of Guadalajara, Zapopan, Mexico

Oswaldo Diaz Big Data Laboratory, INEGI, Aguascalientes, Mexico

Hector A. Duran-Limon Information Systems, CUCEA, University of Guadalajara, Zapopan, Mexico

Lutz Engel Department Management, Information, Technology, Jade University of Applied Sciences, Wilhelmshaven, Germany

Sergio Galvan-Cruz Electronic Systems, Autonomous University of Aguascalientes, Aguascalientes, Mexico

Mick Geisthardt Department of Computing Science, Carl von Ossietzky University of Oldenburg, Oldenburg, Germany;
Department Management, Information, Technology, Jade University of Applied Sciences, Wilhelmshaven, Germany

Jorge Marx Gómez Department of Informatics, Carl von Ossietzky University of Oldenburg, Oldenburg, Niedersachsen, Germany

Martha Hernández-Ochoa Knowledge Fundamentals, CUNORTE, University of Guadalajara, Colotlán, Mexico

Jonas Kallisch OFFIS, Oldenburg, Germany

Ayşenur Karakaya Management Information Systems, Izmir Bakircay University, İzmir, Turkey

Manuel Mora Information Systems, Autonomous University of Aguascalientes, Aguascalientes, Mexico

Serhat Peker Management Information Systems, Izmir Bakircay University, İzmir, Turkey

Gloria Phillips-Wren Sellinger School of Business and Management, Loyola University Maryland, Baltimore, MD, USA

Ali Pişirgen Department of Management and Organization, Karamanoğlu Mehmetbey University, Karaman, Türkiye

Mahesh Raisinghani Texas Woman´S University, Denton, TX, USA

Paola Yuritzy Reyes-Delgado Information Systems, Autonomous University of Aguascalientes, Aguascalientes, Mexico

Leilani Uttenreither University of Maryland Global Campus, College Park, MD, USA

Fen Wang Central Washington University, Ellensburg, WA, USA

Musa Yılmaz Management Information Systems, Izmir Bakircay University, İzmir, Turkey

Chapter 1
A Review of Main Non-proprietary Domain-Independent Data Science Analytics AI/ML Reference Architectures—A Dual ISO/IEC/IEEE 42010 and IT Service Design Approach

Manuel Mora, Paola Yuritzy Reyes-Delgado, Gloria Phillips-Wren, Jorge Marx Gómez, Sergio Galvan-Cruz, and Oswaldo Diaz

Abstract Data Science Analytics and Artificial Intelligence/Machine Learning (DSA-AI/ML) applications aim to create decision-making value by applying analytical procedures from Computer Science, Statistics and AI/ML disciplines to small or big internal and external datasets. However, the development and utilization of applications in this domain requires an adequate DSA-AI/ML architecture. Hence, whereas there is a varied academic literature on DSA-AI/ML Reference Architectures (RAs)—i.e. generic high-level architectural templates for implementing a specific architecture -, such are scarce studies from a dual ISO/IEC/IEEE 42010 Architectural Standard and IT Service Design approach. The ISO/IEC/IEEE 42010 Standard provides valuable guidance for documenting architecture designs and the IT Service Design approach provides best practices to deliver cost-effective IT services. In this chapter, thus, we address this knowledge gap and provide a novel selective-integrative review of main DSA-AI/ML Reference Architectures from a dual ISO/IEC/IEEE 42010 and IT Service Design approach, in the context of non-proprietary

M. Mora (✉) · P. Y. Reyes-Delgado
Information Systems, Autonomous University of Aguascalientes, Aguascalientes, Mexico
e-mail: jose.mora@edu.uaa.mx

G. Phillips-Wren
Sellinger School of Business and Management, Loyola University Maryland, Baltimore, MD, USA

J. M. Gómez
Department of Informatics, Carl von Ossietzky University of Oldenburg, Oldenburg, Niedersachsen, Germany

S. Galvan-Cruz
Electronic Systems, Autonomous University of Aguascalientes, Aguascalientes, Mexico

O. Diaz
Big Data Laboratory, INEGI, Aguascalientes, Mexico

M. Mora et al. (eds.), *Engineering and Management of Data Science, Analytics, and AI/ ML Projects*, Intelligent Systems Reference Library 282,
https://doi.org/10.1007/978-3-032-06889-7_1

domain-independent RAs. Our descriptive review provides theoretical—an updated DSA-AI/ML Architectural Framework—and practical insights—brief but useful descriptions on the purpose and technologies for the DSA-AI/ML Architectural Framework—useful for understanding DSA-AI/ML Reference Architectures.

Keywords Data science analytics AI · ML (DSA-AI/ML) applications · Non-proprietary domain-independent · DSA-AI/ML reference architectures · ISO/IEC/IEEE 42010 standard · IT Service (ITS) design approach

1.1 Introduction

Data Science Analytics and Artificial Intelligence/Machine Learning (DAS-AI/ML) development systems are software systems used to create batch, user interactive or automated DAS-AI/ML applications. Both DAS-AI/ML development systems and application are characterized fundamentally by the utilization of small or big structured and/or non-structured dataset inputs to be processed through data sourcing and collection, data storage and preparation, data modeling and analysis, and data access and utilization phases [1–3]. DAS-AI/ML applications [4] rely on traditional SQL and new Non-SQL retrieval operations, and advanced mathematical, statistical, machine learning, and intelligent heuristics techniques with an Analytics purpose—i.e. descriptive, predictive, or prescriptive—to support business decision making [5].

From a Computer Science perspective, Data Science area was fostered by the availability of huge datasets (i.e. Big Data) [6]. Big Data refers to internal or external huge datasets with 3-V characteristics: very high volume, velocity of generation, and variety of format—[7]. Nowadays, this 3-V scheme has been updated to a 5-V one by adding the veracity and value characteristics [8], and this scheme can be applied for characterizing both small and big datasets [9, 10]. From a Statistics perspective, Data Science area [11] was proposed to rename the Statistics science to Data Science given the three core activities conducted in Statistics: data collection from sample surveys or experimental designs; data modeling and analysis; and problem understanding and solving and decision-making.

Regarding Analytics area, it was initially developed in the Economy science [12] to describe a research method where a problem was formulated with a set of equations, and their analytical and/or approximated solutions were reported including analysis of scenarios during the 1960–2000 period. Later, Analytics area was extended as a relevant approach to Data Science fostered by the convergence of Artificial Intelligence, Machine Learning and Statistics methods used for knowledge data discovery tasks [13]. Nowadays, Analytics can be defined as "the scientific process of transforming data into insight for making better decisions" [14].

DSA-AI/ML applications have contributed to business organizations with benefits such as [15–17]: business process cost reductions, improved quality of decisions, fast correct decision-making, and better visualization of complex datasets, among other business decision-making benefits.

However, to generate such high business impact and value, the development and utilization of DAS-AI/ML applications demands the availability of adequate DSA-AI/ML interrelated Information Technology (IT) resources—of computing, storage, networking, software systems, software applications, data, and IT environment—[18–20]. These interrelated IT resources are organized in an explicitly designed and implemented DSA-AI/ML architecture or an implicit one implemented ad-hoc, and it is known that the functional capabilities and qualities attributes of any IT system are determined by its implemented architecture [21, 22].

To cope with the problematic of selecting a right-sized DSA-AI/ML architecture, there is a varied literature on DSA-AI/ML architectures [23–26]. In the [23, 24] Systematic Literature Reviews (SLR) 19 and 23 final studies were reported. Both initial SLR studies identified main expected architectural quality requirements—Consistency, Scalability, Real-Time Operation, High Performance, Security, Availability, Modularity, and Interoperability—and all of them were mapped to the ISO/IEC 25,010 software quality standard (ISO/IEC, 2011). In [25, 26] SLRs 43 and 22 studies were reported respectively. In the study [25], 40 of the 43 architectures collected were domain-dependent ones—mainly for Social Media, Smart Cities and Healthcare domains –. In the study [26], four architectures were from largest cloud providers (IBM, Microsoft, Oracle, SAP) and two were from standards (NIST BDRA, ISO/IEC 20,547 BDRA). Finally, in all these studies the Lambda architecture was reported as most common architectural base.

Hence, whereas the four SLR [27] studies provides valuable insights on DSA-AI/ML architectures in overall, by using a more quantitative-summative SLR analysis [28], these insights does not cover technical design details for each located DSA-AI/ML architecture neither consider the ISO/IEC/IEEE 42010 architectural Standard [29] and the ITS design approach [30, 31] as lenses of analysis in their reviews. The ISO/IEC/IEEE 42010 Standard provides valuable guidance for documenting architectures and the IT Service Design approach provides best practices to deliver cost-effective IT services. Furthermore DSA-AI/ML development systems and final applications can be provided currently as an IT platform as service (DSA-AI/ML PaaS) [32] or IT software as service (DSA-AI/ML SaaS) [33], and thus the convergence of DSA-AI/ML architectures, the ISO/IEC/IEEE 42010 Standard and the IT Service design approach is relevant for research due to the technical difficulties to define a right-sized DSA-AI/ML architecture. Thus, in this chapter, these later two conceptual schemes are used for conducting an integrative and selective review of main DSA-AI/ML Reference Architectures reported in the literature. Reference Architectures provides generic blueprints usually proposed from a relevant integrative literature in the domain studied [34].

This research chapter updates and extends an initial selective study [35] that reviewed three non-proprietary domain-independent DSA-AI/ML Reference Architectures using a Conceptual Framework based on the pursued dual ISO/IEC/IEEE 42010 architectural Standard and the IT Service Design approach perspectives. Our aim, thus in this research chapter, is to provide and updated and extended useful theoretical and practical integrated set of insights for academic and practitioners interested in the systematic implementation of a non-proprietary domain-independent DSA-AI/ML architecture as an IT service.

This chapter continues as follows. In Sect. 1.2, we describe the Research Approach. In Sect. 1.3, the theoretical foundations of DAS-AI/ML applications, the main elements of the ISO/IEC/IEEE 42010 Standard, and the foundations of IT Service Design approach are reported. In Sect. 1.4, the selected 3 main DSA-AI/ML Reference Architectures are reviewed using the ISO/IEC/IEEE 42010 Standard and the IT Service Design approach conceptualizations as lenses for analysis. In Sect. 1.5, a discussion of theoretical and practical contributions is presented. Finally, in Sect. 1.6 the conclusions, limitations and recommendations for further research are reported.

1.2 Research Approach

This research applies a Selective Conceptual Review (SCR) research methodology adapted from [36, 37], where an integrative-descriptive purpose was pursued. A Systematic Literature Review (SLR) research method [27, 28] was not applied in this research chapter because our research purpose was not to provide a quantitative-oriented accountability of findings on an exhaustive scientific and professional literature search on Data Science Analytics AI/ML Reference Architecture. SLR is usually conducted for mature domains to generate quantitative-based summaries of attributes-topics from the vast generated knowledge rather than to provide deep conceptual reviews on a small but representative group of studies [27, 28]. Because Data Science Analytics AI/ML Reference Architecture research stream is still under developing, we consider worthy to use a SCR research method.

This SCR was performed through four general steps: SCR.1 Research Definition; SCR.2 Research Purpose and Method; SCR.3 Conceptual Data Collection; and SCR.4 Conceptual Analysis and Synthesis. Table 1.1 summarizes the SCR applied in this research chapter. Table 1.2 adds information on the SCR.3 Conceptual Data Collection step. Six RAs emerged as the main ones by its historical relevance, and being frequently cited in the literature, and its available documentation.

These six RAs were: Lambda BDAA [38], RABDS [41], BDAS [39], CCABDA [42], NIST BDRA [40], and the ISO/IEC 20,547 BDRA [43]. RABDS [41] was not selected because it was already analyzed in a previous study of the Research Team [35]. CCABDA [42] was not selected because this qualifies more as proprietary than non-one. The ISO/IEC 20,547 BDRA [43] was considered highly relevant but the Research Team decided including it in a further extension of this research due to its novelty. To manage the extension limitations, then, the Research Team finally selected these three RAs: Lambda BDAA [38], BDAS [39] and NIST BDRA [40]. Lambda BDAA [38] and BDAS [39] can be considered the first RAs reported in the literature. NIST BDRA [40], despite it was already analyzed in [35], it can be

Table 1.1 Selective conceptual review research steps

Step	Purpose	Outcomes	Outcomes in this research
SCR.1. To formulate the research goal	To state the expected research goal indicating the theoretical, or practical or both ones expected contributions	Research goal statement	To contribute to the literature with an updated selective conceptual integrative-descriptive review of the main non-proprietary domain-independent Data Science Analytics AI/ML Reference Architectures from a dual perspective on the ISO/IEC/IEEE 42010 Standard and the IT Service Design approach
SCR.2 To define data sources and selective criteria	To identify and agree the set of data sources to collect the studies, as well as to define the selection criteria	List of data sources Selection criteria statements	The single data source was pre-selected manually. Research Team identified 5 relevant literature review studies [23–26, 35] published in the period 2017–2024. We used these 5 studies as the single data source. To select the specific 3 main Data Science Analytics AI/ML Reference Architectures, we reviewed the list of the Reference Architectures reported in these 5 studies, and the Research Team agreed to select the 3 main reported in most studies
SCR.3 To collect studies	To get the studies	Set of selected studies	From the SCR.2 step, the 3 main studies were [38–40]. [38] refers to Lambda BDA, [39] refers to BDAS, and [40] refers to NIST BDRA v3
SCR.4 To review and synthetize the findings from the collected studies	To conduct the analysis and integration of finding	Structured schema of findings	For the analysis and synthesis of the 5 studies on Data Science Analytics AI/ML Reference Architectures, we used an updated Conceptual Framework reported in [35]. This framework includes conceptual elements of the ISO/IEC/IEEE 42010 Standard and the ITS design approach
(5) To elaborate report of findings	To produce visible results	Research results	This research chapter was elaborated

considered the most cited non-proprietary domain-independent RA for Data Science Analytics AI/ML domain. The other no selected relevant RAs will be analyzed in a further extension of this research.

Table 1.2 Selection of the three main non-proprietary domain-independent Data Science Analytics AI/ML reference architectures

ID study	Year	Main reported non-proprietary domain-independent data science analytics AI/ML reference architectures	Main selected non-proprietary domain-independent data science analytics AI/ML reference architectures
[23]	2017	– RABDS [41]	– None
[24]	2020	– Lambda BDAA [38] – RABDS [41] – NIST BDRA v3 [40]	– Lambda BDAA [38] – NIST BDRA v3 [40]
[25]	2020	– Lambda BDAA [38] – BDAS [39] – RABDS [41]	– Lambda BDAA [38] – BDAS [39]
[26]	2022	– Lambda BDAA [38] – RABDS [41] – NIST BDRA v3 [40] – ISO/IEC 20,547 BDRA [43]	– Lambda BDAA [38] – NIST BDRA v3 [40]
[35]	2022	– NIST BDRA v3 [40] – RABDS [41] – CCABDA [42]	– NIST BDRA v3 [40]

1.3 Background

1.3.1 Foundations of Data Science, Analytics and AI/ML Applications

Data Science Analytics AI/ML applications can be defined as software systems implemented on small or large distributed IT systems configured to manage, process and use internal and/or external datasets with low or high volume, low or high velocity, low or high variety, and high veracity and value for a descriptive-exploratory, predictive or prescriptive business purpose useful for business decision making [1–5].

Volume, Velocity, Variety, Veracity and Value are the 5-V core characteristics that initially characterized only big datasets [8] but currently these 5-V attributes can be applied to small and big datasets [9, 10]. Volume refers to quantity of business events to be registered that demands adequate storage capabilities. Velocity refers to the data generation rate. Variety refers to the rich diversity of data sources (internal vs external, manual user vs automated machine generated, real-time vs batch ingestion engines), data structuredness (SQL vs non-SQL), data formats (text vs binary, analogic vs digital, encrypted vs non-encrypted), and data types (char, string, integer, real, image, sound, video). Veracity refers to the overall quality—characterized by objectivity, truthfulness and credibility—of the datasets. Finally, Value refers to the tangible (cost reductions, profitability increments, business efficiency metrics, among others) and intangible (strategy, business reputation, market value, among others) benefits derived by processing the datasets.

Table 1.3 illustrates the usual range of characteristics between DSA-AI/ML applications processing Small vs Big datasets based on [3].

A full DSA-AI/ML application usually runs a processing workflow—called usually as Data Analytics Pipeline—of several stages. Adapted from [1–3, 44] a Generic DSA-AI/ML application pipeline includes the following stages: (1) Data Sourcing and Collection; (2) Data Storage and Preparation; (3) Data Modeling and Analysis; and (4) Data Access and Utilization.

In the Generic DSA-AI/ML application pipeline concurs two data-based computational approaches. The "Data Science" side refers to the stages of: (1) Data Sourcing and Collection; and (2) Data Storage and Preparation phases. The "Analytics AI/ML" side corresponds to (3) Data Modeling and Analysis; and (4) Data Access and Utilization phases.

Table 1.4 summarizes the stages, purpose, main tasks and key issues for a Generic DSA-AI/ML application pipeline derived and adapted from the several relevant literature [1–3, 44].

To summarize these relevant foundations, the "Data Science" pipeline side is responsible for making available prepared datasets with the potential of creating business value, and the "Analytics AI/ML" pipeline side for providing business value through the application of Analytics and AI/ML procedures to the prepared datasets.

Regarding the (3) Data Modeling and Analysis phase, there are three types of Analytics procedures: Exploratory-Descriptive, Predictive and Prescriptive types. Exploratory-Descriptive Analytics refer to procedures to report summary metrics and graphs of the datasets that represent historical and current status of the business processes related to the datasets. Predictive Analytics refer to procedures to create data-driven models that permit estimate future status of the business processes related to the datasets. Prescriptive Analytics refer to procedures to create data-driven models that determine the optimal solutions or best viable alternative solutions.

1.3.2 Foundations of Systems Architecture Descriptions with the ISO/IEC/IEEE 42010 Standard

The architecture of a system conveys the "*fundamental concepts or properties of a system in its environment embodied in its elements, relationships, and in the principles of its design and evolution*" [29; p. 2]. The architecture of a system, which is an abstraction, is manifested through the functional capabilities and the non-functional properties of the system. Functional capabilities are the specific can-do actions that the system can realize. No-Functional properties—also known as Quality Attributes—are characteristics that the system exhibits as a whole such as performance, reliability, scalability, maintainability, robustness and security among others [29, 45].

Table 1.3 Comparative of DSA-AI/ML applications using small versus big datasets

Attribute	DSA-AI/ML applications using small datasets	DSA-AI/ML applications using big datasets
Data volume	Number of records from thousands to millions. Size datasets from MBs to TBs. Datasets can be stored in a single data server	Number of records from millions to billions or more. Size datasets from TBs to PBs or more. Datasets must be stored in a cluster of data servers
Data variety	Datasets contain structured data (business records). Datasets sources are mainly internal business OLTP and Data Marts systems. Datasets are recorded using SQL and relational technologies	Datasets contain more semi-structured and unstructured (xml/json texts, text documents, binary images, binary audios, binary videos, binary streams, sensor signals) than structured data (business records). Datasets are recorded using No SQL, graph, key-value, and columnar technologies
Data velocity	Rates of generated-collected business records are in the range of hundreds to thousands per hour. Most final processed data uses batch mode	Rates of generated-collected unstructured items are in the range of thousands to millions per hour. Most final processed data uses real-time mode
Data veracity	Very high data quality due to the main utilization of structured internal business data sources, and the utilization of an ETL process	From moderate to high data quality due to the main utilization of unstructured external business data sources. An ELT is used
Data value	There is an explicit and current utility value due to the need to count with the datasets for supporting business processes	There is an implicit and potential utility value due to there is not mandatory to count with the datasets for supporting business processes
IT resources	Usually a centralized single or a small processing-storage server cluster	Moderate to large, distributed processing-storage server cluster
IT people	High-skilled on analytics	High-skilled on analytics, big data science, and big data IT services
Development process	Mainly CRISP-DM or a business intelligence dashboard designs	There is not a specific and standard development methodology
Development cost	Affordable costs (development team, development IT resources, structured datasets)	High costs (development team, development IT resources, unstructured datasets)
System lifespan	SDAS usually are used for long time spans—years—due to the stability of datasets, functional requirements and used data processing techniques	BDAS usually are used for short time spans—months—due to the dynamism of datasets, functional requirements and used data processing techniques

(continued)

Table 1.3 (continued)

Attribute	DSA-AI/ML applications using small datasets	DSA-AI/ML applications using big datasets
Analytics purpose	Mainly used for descriptive/exploratory and predictive purposes. Prescriptive purposes are also pursued	Firstly used for descriptive/exploratory purpose. Secondly used for predictive purpose
End users	Business managers/analysts	Big data business managers/analysts

According to [29], the architecture of a system can be designed and represented through an Architecture Description (AD) document. An AD document reports stakeholders and their concerns, architecture decisions and rationale, and architecture views and viewpoints. Stakeholders are any entity that will be affected by the system of interest. Concerns are the expected system properties of interest for the stakeholders. Architecture decisions and rationale are the architectural design selections done and their justifications. Architecture views are diagrammatic representations—called architecture models—governed by architecture viewpoints that depicts a set of specific concerns. Relevant literature on architecture design adds architectural view types and styles [45]. View types are pre-defined views and styles are pre-defined particular design representations into a view type. In [45] there are 4 predefined view types: Module, Component-and-Connector, Allocation, and Hybrid view types. In the Module view type the main design styles are Decomposition-Functional and Layered-Functional ones. In the Component-and-Connector view type the main design styles are Client–Server and SOA. In the Allocation view type the main design style is Deployment one.

Consequently, an AD usually includes this set of terms: architecture description content, stakeholders and their architectural concerns, architectural functional capabilities, architectural quality attributes, architecture design decisions and their rationale, architectural views, view types, design styles and their viewpoints.

To guide systems architects in the design of a system architecture, have been proposed Architecture Frameworks (AF), Reference Architectures (RA), and Architecture Design Processes and Practices (ADPP) [29, 34]. An AF "*establishes a common practice for creating, interpreting, analyzing and using architecture descriptions within a particular domain of application or stakeholder community*" [29; p. 9]. A RA refers to "*a generic architecture for a class of systems that is used as a foundation for the design of concrete architectures from this class*" [34; p. 417]. ADPP define the activities and practices for: (1) analyzing the functional and non-functional architectural requirements, (2) designing candidate architectures, and (3) evaluating and selecting the solution architecture.

AFs and RAs applied directly to IT services are scarce in the IT Service Design literature [30, 31]. IT4IT [44] and TOGAF [46] can be reported as the main relevant but they are focused on IT function [44] or Enterprise Architecture [46] rather on IT Service Design. Nevertheless, the main IT Service Design literature [30, 31] have

Table 1.4 A generic DSA-AI/ML application pipeline

Pipeline stage	1. Data sourcing and collection	2. Data storage and preparation	3. Data modeling and analysis	4. Data access and utilization
Purpose	To identify the set of raw data sources for the DSA-AI/ML application pipeline, agree legally on its accessibility, collect the agreed raw data, transmit them, and register them	To apply data processing procedures to raw datasets, to pull datasets of interest, to apply data preparation procedures, and to load them in the persistent storage platforms	To elaborate data-driven models and apply those analytics procedures for specific business goals	To use data-driven models in batch, interactive and/or real-time automatic embedded modes for specific business goals
Main tasks	1.1 Identification, analysis and selection of the available raw data sources 1.2 Raw data collection, transmission and ingestion 1.3 Raw data catalog registering	2.1 Raw data pre-processing (compression/ decompression, cleaning, and redundancy elimination) 2.2 Cleaned data transformation, integration, aggregation, and representation 2.3 Cleaned and prepared data storage and replication	3.1 Exploratory and descriptive analytics (OLAP, descriptive statistics, descriptive charts/graphs) 3.2 Predictive analytics (classification, regression, clustering, association) 3.3 Prescriptive analytics (optimization, simulation, heuristic methods, expert systems)	4.1 Access and use of DSA-AI/ML batch applications 4.2 Access and use of DSA-AI/ML interactive applications 4.3 Access and use of DSA-AI/ML real-time automatic embedded applications
Key issues	Variety of raw data formats (structured, text, image, audio, video, device signal). Velocity (generation rates of raw data). Volume (raw data size). Veracity (trust level of raw data). Value (business need for raw data). QoS metrics for LAN/ WAN/Internet data transmission systems	Performance metrics for data storage and processing preparation platforms. Data security and privacy issues	Performance metrics for analytics processing platforms. Taxonomy of exploratory-descriptive, predictive and/or prescriptive analytics procedures	QoS metrics for LAN/ WAN/Internet data transmission systems. Usability metrics. Performance metrics. Business goals metrics

provided the best processes-practices to deliver business value to IT users through the concept of IT services.

1.3.3 Foundations of ITS Design Approach

According to the ITS Design Approach literature [30, 31], and IT Service can be defined as a functionality enabled to IT users that generates business value and it is delivered by an IT Service System composed of IT resources, IT processes and practices, and IT people. Figure 1.1—adapted from [30, 31]—illustrates the concept of IT Service and IT Service System.

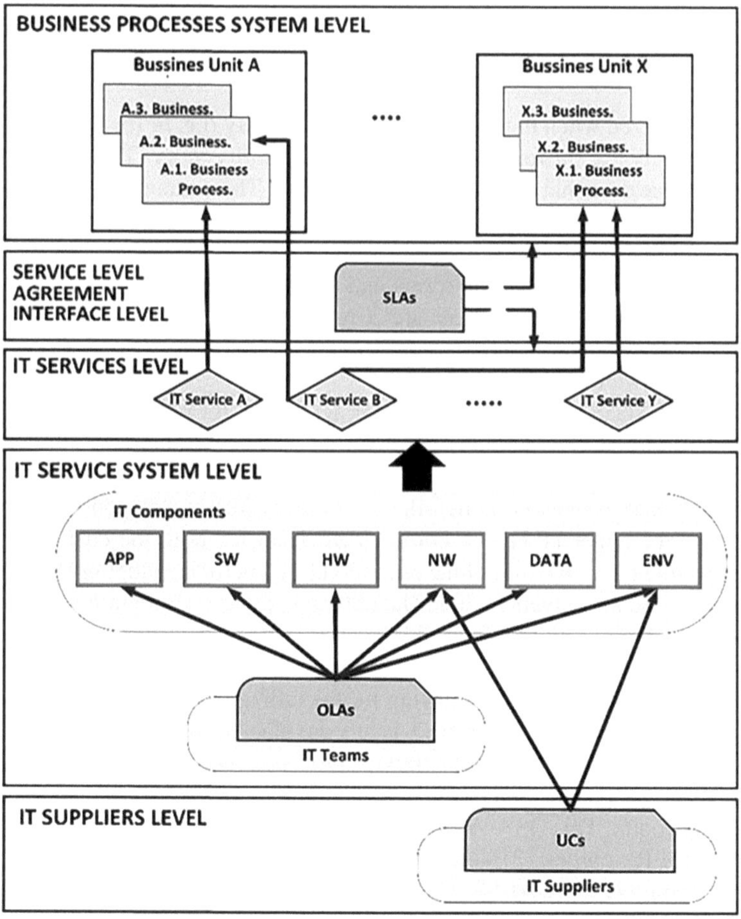

Fig. 1.1 IT service and IT service system concepts

Figure 1.1 shows 5 levels for representing an IT Service: Business Processes System level, Service Level Agreement (SLA) Interface level, IT Service level, IT Service System level, and IT Suppliers level. The first level accounts for the set of business processes that can be supported by one or more IT Services agreed with an SLA. Second level refers to the set of SLAs that establish the expected IT Service functionalities, metrics, schedule and other relevant information on the agreed IT Services for the IT customers. Third level represents the IT Services provisioned to the first level—under agreed SLAs -. The fourth level accounts for the system that generates and provides the IT Service functionalities as well as pursues to achieve the agreed IT Service metrics. This level includes the IT components of software applications (APP), software base (SW), hardware equipment (HW), network equipment (NW), data (DAT), all environmental equipment and facilities (ENV), and IT service teams. This fourth level also includes the Operational Level Agreements (OLAs) agreed internally in the diverse IT Service teams and the Underpinning Contracts (UCs) that are agreed between the IT Service organization and the external IT providers required to provide IT Services. Last fifth level represents these external IT suppliers.

Value is realized when the expected IT Service utility (i.e. fit for purpose) and IT Service warranty (i.e. fit for use) is achieved. The utility of an IT Service refers to what the service does that is valued by the customer. The warranty for an IT Service refers to how well it is delivered—i.e. how well are reached the levels of availability, capacity, continuity, and security agreed -.

DSA-AI/ML development platforms and applications can be delivered as an IT Service [32, 33]. In general, there are 3 provision IT Service modes: DSA-AI/ML Infrastructure as Service (IaaS), development Platforms as Service (IaaS), and application Software as Service (SaaS).

DSA-AI/ML IaaS refers to the customer agreement for paying the utilization of physical and virtual IT resources. The IaaS provider owns and hosts the physical IT resource layer, but the DSA-AI/ML customer remotely manages them. In this DSA-AI/ML IaaS provision model, the customer is free and responsible to install and manage the upper IT layers. DSA-AI/ML PaaS refers to the customer agreement for paying the utilization of the required IT layers for developing DSA-AI/ML applications. Usual IT layers include Data Storage Servers, Computing Servers, and Development Tools. The two lower IT layers are considered black boxes, and the next upper IT layer is the responsibility of the customer. Finally, DSA-AI/ML SaaS refers to the customer agreement paying for the utilization of an end-user DSA-AI/ML application software. All lower IT layers are black boxes for the customer.

Figure 1.2 illustrates the three IT Service provision modes for DSA-AI/ML as IT Services using a Layered-Functional Architectural view [29, 45] from an IT Service provider viewpoint. This DSA-AI/ML Layered-Functional architecture has 6 layers: (1) Physical IT Resources; (2) Virtual IT Resources; (3) DSA-AI/ML Storage Servers Administration; (4) DSA-AI/ML Computing Servers Administration; (5) DSA-AI/ML Development Tools; and (6) DSA-AI/ML Applications. Figure 1.2 maps also the Generic DSA-AI/ML Application Pipeline reported in Table 1.3 to the 3 ITS provision modes.

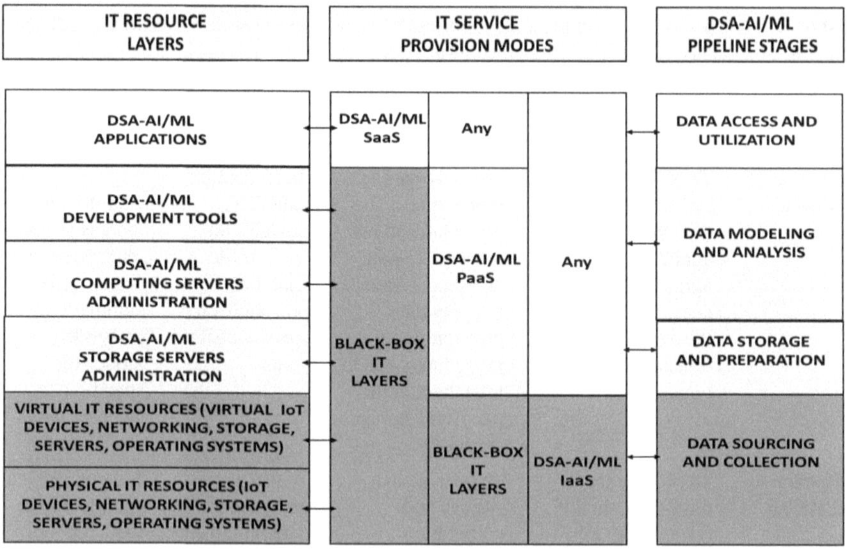

Fig. 1.2 Types of provision modes for DSA-AI/ML as IT services using a layered-functional architectural view

Finally, Table 1.5 combines the Generic DSA-AI/ML Application Pipeline reported in Table 1.4 with the Layered-Functional Architectural view illustrated in the Fig. 1.2. Pipeline stages were mapped to the Layer-Functional level based in the utilization of the latter on the former ones. For instance, the Data Access and Utilization pipeline stage is performed using the elements existent in the DSA-AI/ML layer-function. Similarly, the Data Sourcing and Collection pipeline stage can be executed by using the elements in the Virtual and Physical IT Resources layer-function.

1.4 Selective Review of the Main DSA-AI/ML Reference Architectures with an Integrative-Descriptive Purpose

1.4.1 The DSA-AI/ML Architectural Framework as Conceptual Lenses

To conduct this Integrative-Descriptive review, we use -as conceptual lenses for analysis—the DSA-AI/ML Architectural Framework, which includes the Generic DSA-AI/ML Application Pipeline (Table 1.4) and it is represented in the Layered-Functional Architectural view (Fig. 1.2).

This Layered-Functional Architectural view (Fig. 1.2) includes 6 functional layers: (1) Physical IT Resources; (2) Virtual IT Resources; (3) DSA-AI/ML Storage

Table 1.5 A DSA-AI/ML architectural framework—including generic DSA-AI/ML application pipeline and the layered-functional architectural view

Pipeline stage	1. Data sourcing and collection	2. Data storage and preparation	3. Data modeling and analysis	4. Data access and utilization
Pipeline purpose	To identify the set of raw data sources for the DSA-AI/ML application pipeline, agree legally on its accessibility, collect the agreed raw data, transmit them, and register them	To apply data processing procedures to raw datasets, to pull datasets of interest, to apply data preparation procedures, and to load them in the persistent storage platforms	To elaborate data-driven models and apply those analytics procedures for specific business goals	To use data-driven models in batch, interactive and/or real-time automatic embedded modes for specific business goals
Pipeline function	To provide the physical/virtual IT resources to DSA-AI/ML to collect, transmit and ingest raw data, as well as for upper pipeline layer functions	To provide data storage and preparation processing computational capabilities	To provide data analytics computational engines for batch, interactive, and/or real-time automatic embedded processing modes	To provide data computational tools for accessing and usage in batch, interactive and/or real-time automatic embedded modes
Type of IT Service	DSA-AI/ML IaaS	DSA-AI/ML PaaS	DSA-AI/ML PaaS	DSA-AI/ML SaaS
IT resources layers	Physical and virtual IT resource layers	Storage servers administration layer	Development tools and computing servers administration layer	Applications layer
IT resources categories	Virtualization platforms Cloud platforms Processing servers clusters Real-time streaming engines	Storage servers clusters Data lakes platforms Small and big data SQL and non-SQL databases Small and big data warehouses Small and big data OLAP engines	Small and big data analytics platforms Small and big data AI/ML platforms Small and big data graph analytics platforms	Batch small and big data visualization analytics platforms Interactive small and big data visualization analytics platforms Real-time small and big data visualization analytics platforms

Servers; (4) DSA-AI/ML Computing Servers; (5) DSA-AI/ML Development Tools; and (6) DSA-AI/ML Applications. The 2 bottom layers correspond to the DSA-AI/ML IaaS provision mode. The next 3 layers correspond to the DSA-AI/ML PaaS and the last top layer corresponds to the DSA-AI/ML SaaS provision modes.

In this Layered-Functional Architectural view, it is included the Generic DSA-AI/ML Application Pipeline with 4 stages. The first stage of Data Sourcing and Collection is mapped to the 2 bottom layers of Physical and Virtual IT resources, and it corresponds to an IaaS provision mode. Internal and external, structured and non-structured, and batch, interactive or stream data sources from business enterprise systems, business devices, external IoT networks, social networks, external open data repositories, and external commercial data repositories need to be identified and acquired. LAN/WAN/Internet data transmission systems, cloud platforms, real-time streaming engines (such as Kafka, Flink, or Storm), and IoT sensors databases (such as IoTDB), are IT resources categories of design components that must be also considered for the first stage.

The second stage Data Storage and Preparation was mapped to the third layer of DSA-AI/ML Storage Servers. This layer refers to the Small and Big Data tools for managing the SQL- and non-SQL based storage through a cluster of storage servers, as well as for pre-processing (compression/decompression, cleaning, redundancy elimination, transformation) and preparation (integration, aggregation, representation, replication, and processed data ingestion) tasks. In this third layer, the IT resources categories of design components are IT cluster management systems (such as Mesos, YARN), Small and Big Data pre-processing tools (such as CKAN, Apache Griffin, Open Refine, DataCleaner), Storage Servers clusters (such as Hadoop/HDFS), Storage Processing Engines (such as Apache Pig), Small and Big Data warehouses (such as Hive, Impala, BigQuery, Presto), and Small and Big Data non-SQL databases (such as MongoDB, Cassandra, HBase).

The third stage of Data Modeling and Analysis was mapped to the fourth and fifth layers of DSA-AI/ML Computing Servers and DSA-AI/ML Development Tools. These layers enable the design and building of data-driven models and the application of analytics procedures for specific business goals. Analytics procedures can be Exploratory and Descriptive (e.g. OLAP, descriptive statistics, and descriptive charts/graphs), Predictive (e.g. classification, regression, clustering, and association), and Prescriptive (e.g. optimization, simulation, heuristic methods, and expert systems). In these fourth and fifth layers, the IT resources categories of design components are Analytics Servers clusters, Small and Big Analytics engines (such as Mahout, Apache Drill, Spark, MLlib, RHadoop, RHive, TensorFlow, Pytorch, Keras), and Small and Big Graphs engines (such as GraphX, GraphLab, neo4j, Giraph, ArangoDB). These 3 layers correspond to the PaaS provision mode.

The final fourth stage of Data Access and Utilization was mapped to the sixth layer of DSA-AI/ML Applications. This top layer enacts the local and remote access and utilization of the data-driven models in batch, interactive or real-time streaming modes for specific business goals. This sixth layer corresponds to the SaaS provision mode.

1.4.2 Review of the Main Three DSA-AI/ML Reference Architectures

As it is reported in the Sect. 1.2, through the applied Selective Review, three main DSA-AI/ML Reference Architectures were selected: Lambda BDAA [38], BDAS [39], and NIST BDRA [40].

Lamda BDAA—Big Data Analytics Architecture—[38] was designed to cope with the challenges on DSA-AI/ML architectures for keeping adequate performance and robustness and manageable complexity scalability when required. Lamda BDAA designers [38] indicated that the unmanaged utilization of the myriad of DSA-AI/ML tools and platforms did not provide adequate Ad-Hoc architectural solutions, and thus a planned DSA-AI/ML architecture is required. Lamda BDAA [38] proposes an RA with three layers: (1) Batch layer; (2) Serving layer; and (3) Speed layer. Batch layer is responsible for keeping an immutable master dataset and generate batch data views pre-computed periodically when new data packages of predefined cardinality are received in the DSA-AI/ML architecture. The immutable master dataset means that new data can be added but current already stored data cannot be deleted or changed. Serving layer is responsible to execute low-latency random-access reads—but not random-access writes -on the periodically pre-computed batch views. These batch views are updated periodically. Finally, Speed layer is responsible for creating real-time data views through incremental updates for new data packages that were not included in the last elaboration of batch views. Speed layer requires random-access reads and writes but its latency is adequate due to the smaller size of the real-time data views. Lamda BDAA [38] exhibits an implicit high-level DSA-AI/ML pipeline that can be defined as: (1) Batch and Real-Time Data Ingestion stage; (2) Batch and Real-Time Data Views Preparation stage; and (3) Interactive and Real-Time Data Query Utilization stage. Implicit Analytics capabilities are focused on Descriptive type. References to ISO/IEC/IEEE 42010 conceptualizations are null but anyway Lambda BDAA [38] reports: architectural functional capabilities, architectural quality attributes, architecture design decisions and their rationale, and an architectural view—Module type with a Layered-Functional style -. References to ITS design components and provision modes are also omitted. Figure 1.3 exhibits the Lambda BDAA mapped to the DSA-AI/ML Layered-Functional Architectural View.

BDAS (Berkely Data Analysis System) [39], was proposed as an Architectural Stack for Big Data Analytics supported by Open Source IT. BDAS started as a sponsored military project with the technical challenge to have a Big Data Analytics platform that could offer strong processing capabilities—under time, cost and quality constraints -. This challenge was identified given the disparity between the data generation and collection capabilities versus data processing ones. The added requirement of using Open Source IT was for its public dissemination and utilization.

BDAS [39] has five layers: (1) Resource Virtualization layer; (2) Storage layer; (3) Processing Engine layer; (4) Access and Interfaces layer; and (5) Applications layer. First Resource Virtualization layer administrates a cluster of physical servers through

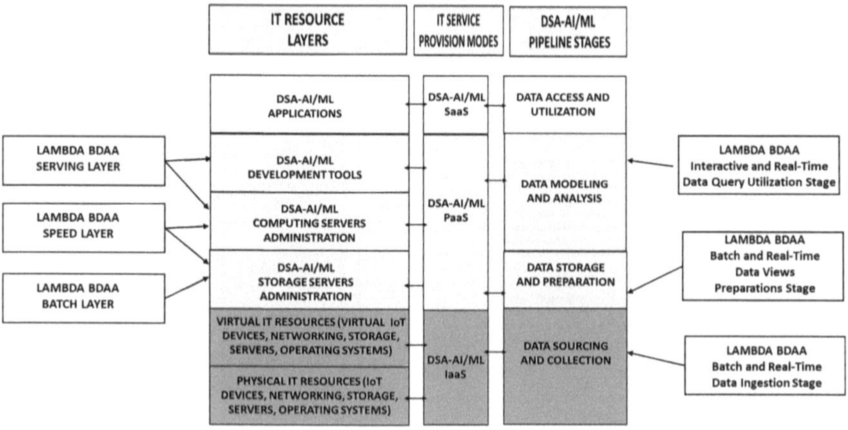

Fig. 1.3 Lambda BDAA mapped to the DSA-AI/ML layered-functional architectural view

Mesos Cluster system, which abstracts physical IT resources to be used by upper layers. Second Storage layer administrates a memory-centric, fault-tolerant virtual distributed storage system through the Alluxio system. This layer can also manage a traditional disk-based Hadoop distributed storage system. Third Processing Engine layer provides a fast in-memory computing cluster system to read, process and write data through Apache Spark system. The fourth Access and Interfaces layer provides a set of complementary capabilities to core computing ones. These complementary capabilities refer mainly to: data streaming processing, SQL processing, big graph processing, and ML procedures and pipelines. Finally, the last fifth Applications layer refers to the end-user applications served by the previous layers.

BDSA [39], similar to BDAA [38] exhibits an implicit high-level DSA-AI/ML pipeline that can be defined as: (1) Batch, Interactive and Streaming Data Ingestion and Preparation stage; (2) Batch, Interactive and Streaming Data Processing stage; and (3) Batch, Interactive and Streaming Data Utilization stage. Implicit Analytics capabilities are focused on Descriptive and Predictive types. References to ISO/IEC/IEEE 42010 conceptualizations are null but anyway BDAS [39] reports architectural functional capabilities, architectural quality attributes, architecture design decisions and their rationale, and an architectural view—Module type with a Layered-Functional style -. References to IT Service Design components and provision modes are also omitted. Figure 1.4 exhibits the BDAS mapped to the DSA-AI/ML Layered-Functional Architectural View.

NIST BDRA (Big Data Reference Architecture) [40] was designed by NITS (National Institute of Standards and Technology, USA) after several rounds of sessions in the NIST Big Data Public Working Group (NBD-PWG) with participants from industry, academia, and government agencies. According to NIST [40; p. 3] a Reference Architecture provides "an authoritative source of information about a specific subject area that guides and constrains the instantiations of multiple architectures and solutions." NIST BDRA [40] is reported as vendor-neutral technology- and

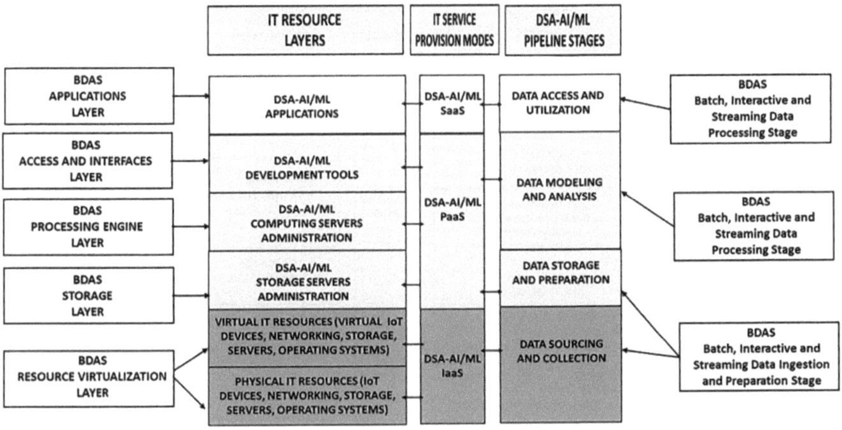

Fig. 1.4 BDAS mapped to the DSA-AI/ML layered-functional architectural view

infrastructure-agnostic conceptual model and two architectural views (Activity View and Functional view). NIST BDRA [40] pursues to reach adequate quality architectural attributes of interoperability, portability, reusability, extensibility, security and privacy, and functional ones of data usage, analytics processing, and technology infrastructure management.

NIST BDRA [40] is designed with five IT layers—named as Role layers—all supported with two fabric IT roles and five DSA-AI/ML pipeline stages. The five IT layers are: (1) Big Data Provider Framework with 3 IT sub-layers (Networking, Computing and Storage Infrastructures sub-layer; Data Organization and Distribution Platforms sub-layer; and Computing and Analytics Processing sub-layer); (2) Big Data Application Provider; (3) System Orchestrator; (4) Data Provider; and (5) Data Consumer. The two fabric IT roles are Security and Privacy fabric and Management fabric, and these ones provide critical internal support services for the core 5 IT layers.

The first Big Data Provider Framework layer provides: (1) the physical and virtualized IT resources of networking, computing and storage; (2) the capabilities for organizing and managing distributed data; and (3) the core batch, interactive and streaming computing and analytics capabilities. The second Big Data Application Provider layer provides the functionalities of a generic DSA-AI/ML pipeline of 5 stages from end-users perspective: (1) Data Collection; (2) Data Preparation; (3) Data Analytics; (4) Data Visualization; and (5) Data Access. The third System Orchestrator layer refers to several high-level managerial capabilities to operate the other IT layers. These managerial capabilities are such as: governance requirements and monitoring, business ownership requirements and monitoring, data science requirements and monitoring, system architecture requirements definition and security/privacy requirements definition and monitoring. The fourth Data Provider layer is responsible for controlling the data ingestion, persistent registering and transformation/preparation and push datasets to the Big Data Application Provider layer or permit pull datasets

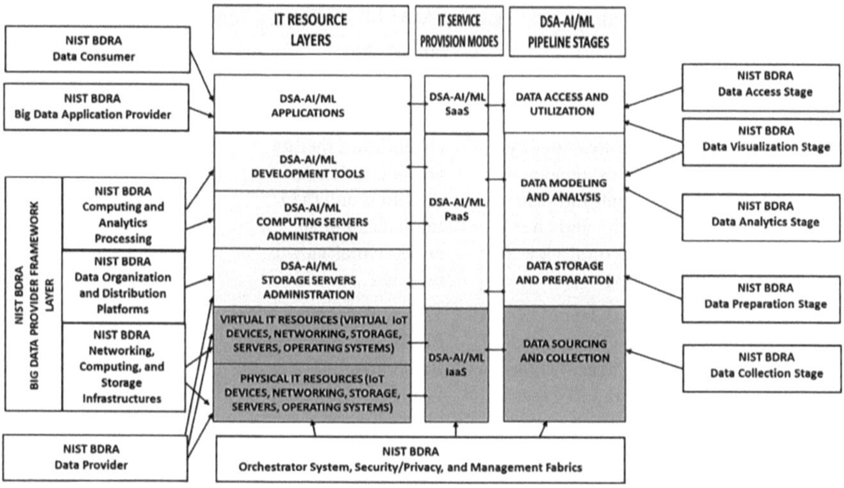

Fig. 1.5 NIST BDRA mapped to the DSA-AI/ML layered-functional architectural view

from this layer. The fifth Data Consumer layer refers to the functional capabilities for data searching, retrieving, using, reporting, visualizing and transferring to other end-user systems.

References to ISO/IEC/IEEE 42010 conceptualizations are existent through the definition of architectural functional capabilities, architectural quality attributes, architecture design decisions and their rationale, and two architectural views—Activity and Functional ones using Module type with a Layered-Functional style -. References to ITS design components and provision modes are also used. Figure 1.5 exhibits the NIST BDRA [40] mapped to the DSA-AI/ML Layered-Functional Architectural View.

Table 1.6 reports a summary of the main findings found in this research for the three main DSA-AI/ML Reference Architectures analyzed.

1.5 Discussion of Contributions

In this chapter, we have conducted a selective-integrative review on the landscape of the main DSA-AI/ML Reference Architectures from dual ISO/IEC/IEEE 42010 and IT Service Design perspective.

DSA-AI/ML Reference Architectures can be useful for an IT Service Design architect in small, medium and large organizations. In particular, the small ones are benefited due that these kind of organizations usually lack sufficient economic, technical and organizational resources to implement and use DSA-AI/ML as ITS. Consequently, this research chapter provides useful theoretical and practical insights as architectural blueprints for designing and lately implementing a DSA-AI/ML

Table 1.6 Summary of Findings on 3 main DSA-AI/ML Reference Architectures analyzed

Issue	BDAA [38]	BDAS [39]	NIST BDRA [40]
ISO/IEC/IEEE 42010 terminology?	**No.** However, architectural functional capabilities, quality attributes, design decisions and their rationale, and an architectural view are reported	**No.** However, architectural functional capabilities, quality attributes, design decisions and their rationale, and an architectural view is reported	**Yes.** Architectural functional capabilities, quality attributes, design decisions and their rationale, and diverse views and viewpoints are reported
DSA-AI/ML ITS layers?	**Yes.** A 3-layer DSA-AI/ML RA is reported: (1) Batch layer; (2) Serving layer; and (3) Speed layer	**Yes.** A 3-layer DSA-AI/ML RA is reported: (1) Batch layer; (2) Serving layer; and (3) Speed layer	**Yes.** A 3-layer DSA-AI/ML RA is reported: (1) Batch layer; (2) Serving layer; and (3) Speed layer
DSA-AI/ML pipeline stages?	**No.** However, an implicit 3-stage high-level DSA-AI/ML pipeline can be identified	**No.** However, an implicit 3-stage high-level DSA-AI/ML pipeline can be identified	**Yes.** A 5-stage DSA-AI/ML pipeline is reported in the Big Data Application Provider layer
DSA-AI/ML ITS provision modes?	**No.** Only is reported briefly the IaaS provision mode	**No.** ITS provision modes are not reported	**Yes.** ITS provision modes are reported
ITS design terminology?	**Yes.** ITS design terminology is minimally used but the concept of service is supported	**No.** ITS design terminology is not used	**Yes.** ITS design terminology is used
DSA-AI/ML technologies for design components?	**Yes.** Specific DSA-AI/ML technologies are considered for each ITS layer	**Yes.** Specific DSA-AI/ML technologies are considered for each ITS layer	**No.** Specific DSA-AI/ML technologies are not considered for each ITS layer
Contribution to ITS designers?	**Yes.** It provides a well-detailed RA for DSA-AI/ML	**Yes.** It provides a well-detailed RA for DSA-AI/ML	**Yes.** It provides a full comprehensive RA for DSA-AI/ML
Contribution to DSA-AI/ML architectural literature?	**Yes.** It provides an affordable RA for DSA-AI/ML well-documented using implicit architectural terminology	**Yes.** It provides an affordable RA for DSA-AI/ML well-documented using implicit architectural terminology	**Yes.** It provides a RA for DSA-AI/ML well-documented using terminology of the ISO/IEC/IEEE 42010 standard

architecture. The design and implementation of a DSA-AI/ML architecture demands usually high-costs of IT infrastructure or cloud-based fees, of licenses of proprietary DSA-AI/ML software tools, of specialized training, and of high-tech consulting. The availability of Open Source IT tools for implementing a DSA-AI/ML architecture that use moderate-cost IT infrastructure (on-premise or a cloud-based) is an economic affordable and technically feasible path for small organizations.

This research chapter identified three main DSA-AI/ML Reference Architectures (RA): Lambda BDAA [38], BDAS [39], and NIST BDRA [40]. These three DSA-AI/ML RA were selected from five relevant literature review studies selected manually [23–26, 35] that were published in the period 2017–2024. We used these five studies as the single data source. Lambda BDAA [38] can be considered one of the first DSA-AI/ML RA reported in the literature. BDAS [39] also can be considered one of the first DSA-AI/ML RA whose design is based on in-memory dataset processing, accelerating computing and write processing times. NIST BDRA, in turn, can be reported as one of the first DSA-AI/ML RA provided as a standard.

To analyze the three selected DSA-AI/ML RAs, we reported in this research chapter: (1) a comparative profile between DSA-AI/ML Applications using Small vs Big Datasets (Table 1.2); (2) a generic 5-Stage DSA-AI/ML pipeline including main tasks and technical issues per stage (Table 1.3); (3) a brief but succinct description of the architectural foundations for IT Service Design from the ISO/IEC/IEEE 42010 standard and the related literature; (4) a description of the modern concept of IT service, including the three IT Service provision modes, and its architectural functional view (Fig. 1.1); and (5) a DSA-AI/ML Architectural Framework of IT Services documented with a Layered-Functional Architectural View (Fig. 1.2 and Table 1.4). This DSA-AI/ML Architectural Framework is composed of six IT Service layers and includes a 5-Stage pipeline.

The six IT Service layers are: (1) Physical IT Resources; (2) Virtual IT Resources; (3) DSA-AI/ML Storage Servers Administration; (4) DSA-AI/ML Computing Servers Administration; (5) DSA-AI/ML Development Tools; and (6) DSA-AI/ML Applications. The 5 pipeline stages are: (1) Data Sourcing and Collection; (2) Data Storage and Preparation; (3) Data Modeling and Analysis; and (4) Data Access and Utilization.

Then, by using this DSA-AI/ML Architectural Framework, the three main DSA-AI/ML RAs were analyzed. Table 1.5 reported a summary of the key findings. Essentially, only the NIST BDRA [40] endorses adequately the architectural and IT Service Design literatures, and reports explicitly architectural IT Service layers, DSA-AI/ML pipeline stages and several architectural issues. However, the other two DSA-AI/ML RAs analyzed—Lambda BDAA [38] and BDAS [39] reported the utilization of architectural concepts and IT Service layers but no a DSA-AI/ML generic pipeline. In contrast, these later two DSA-AI/ML RAs reported specific IT Services technologies to be used in each IT Service layer.

Hence, we consider this research contributes to the literature with the following items: (1) it reports a comparative profile of DSA-AI/ML Applications using Small Datasets vs Big Datasets; (2) it presents a generic 5-Stage DSA-AI/ML pipeline; and (3) it proposes a DSA-AI/ML Architectural Framework—including the generic 5-Stage pipeline—useful to analyze DSA-AI/ML RAs from an IT Service design perspective. This research also contributes to the practice because: (1) it helps to IT Service designers from small organizations to a better understanding of the technical landscape of DSA-AI/ML as an IT Service; (2) it makes explicit a 5-Stage DSA-AI/ML pipeline; and (3) it exhibits an essential DSA-AI/ML Architectural

Framework useful for guiding IT Service designers of small organizations to identify the most adequate and technically viable type of IT Service (as software, platform or infrastructure) of DSA-AI/ML that the business organization can implement.

1.6 Conclusions

This research chapter reviewed three of the main DSA-AI/ML Reference Architectures by using a DSA-AI/ML Architectural Framework from an IT Service Design perspective combined with the ISO/IEC/IEEE 42010 architectural standard and related literature. These DSA-AI/ML RAs were Lambda BDAA, BDAS, and NIST BDRA. From a theoretical perspective only NIST BDRA is more fundamentally supported on the architectural and IT Service design literatures and more well-described than other two ones. However, the other two DSA-AI/ML RAs (Lamda BDAA and BDAS) implicitly report also adequate architectural issues and partially IT Service design core concepts. From a practical perspective, the three DSA-AI/ML RAs provide useful insights to design and implement an economic affordable and technical feasible DSA-AI/ML architecture.

To advance this research stream, we recommend the following avenues for further research: (1) to extend the selection of the main DSA-AI/ML RAs reported in the literature; (2) to refine the analyses conducted on each identified DSA-AI/ML RA; (3) to add a complementary directory of useful IT tools for implementing a DSA-AI/ML architecture; (4) to refine the DSA-AI/ML Architectural Framework; and (5) to implement a DSA-AI/ML architecture and conduct empirical research on its IT Service Design issues.

Hence, further conceptual and empirical research are encouraged.

References

1. Watson, H.J.: Update tutorial: big data analytics: concepts, technology, and applications. Commun. Assoc. Inf. Syst. **44**(1), 364–379 (2019)
2. Phillips-Wren, G., Lakshmi, S.I., Uday, K., Ariyachandra, T.: Business analytics in the context of big data: a roadmap for research. Commun. Assoc. Inf. Syst. **37**(1), 448–472 (2015)
3. Mora, M., Reyes-Delgado, P. Y., Galvan-Cruz, S., Solano-Romo, L. I.: Open source IT for delivering big data analytics systems as services: a selective review. In: Development Methodologies for Big Data Analytics Systems: Plan-driven, Agile, Hybrid, Lightweight Approaches, pp. 1–43 (2023)
4. Kune, R., Konugurthi, P.K., Agarwal, A., Chillarige, R.R., Buyya, R.: The anatomy of big data computing. Softw. Pract. Exper. **46**(1), 79–105 (2016)
5. Delen, D., Ram, S.: Research challenges and opportunities in business analytics. J. Bus. Anal. **1**(1), 2–12 (2018)
6. Cox, M., Ellsworth, D.: Managing big data for scientific visualization. In: ACM SIGGRAPH Proceedings, Los Angeles, CA, 3–8 August 1997 (1997)
7. Laney, D.: 3D data management: controlling data volume, velocity and variety. META Group Research Note 6 (2001)

8. Wamba, S., Akter, S., Edwards, A., Chopin, G., Gnanzou, D.: How 'big data' can make big impact: findings from a systematic review and a longitudinal case study. Int. J. Prod. Econ. **165**, 234–246 (2015)
9. Cao, L.: Data science: challenges and directions. Commun. ACM **60**(8), 59–68 (2017)
10. Sarker, I.H.: Data science and analytics: an overview from data-driven smart computing, decision-making and applications perspective. SN Comput. Sci. **2**(5), 377 (2021)
11. Wu, J.: Statistics= data science (1997). http://www2.isye.gatech.edu/~jeffwu/presentations/datascience.pdf
12. Tedford, J.R.: Analytics of decision making. J. Farm Econ. **46**(5), 1353–1362 (1964)
13. Fayyad, U., Piatetsky-Shapiro, G., Smyth, P.: From data mining to knowledge discovery in databases. AI Mag. **17**(3), 37–37 (1996)
14. INFORMS: Best definition of analytics (2019). https://www.informs.org/About-INFORMS/News-Room/O.R.-and-Analytics-in-the-News/Best-definition-of-analytics
15. Davenport, T.H.: Competing on analytics. Harv. Bus. Rev. **84**(1), 98–107 (2006)
16. Dong, J.Q., Yang, C.: Business value of big data analytics: a systems-theoretic approach and empirical test. Inf. Manag. **57**, 103124 (2020)
17. Klee, S., Janson, A., Leimeister, M.: How data analytics competencies can foster business value–a systematic review and way forward. Inf. Syst. Manag. **38**(3), 200–217 (2021)
18. Oussous, A., Benjelloun, F.Z., Lahcen, A.A., Belfkih, S.: Big data technologies: a survey. J. King Saud Univ. Comput. Inf. Sci. **30**(4), 431–448 (2018)
19. Rao, T., Mitra, P., Bhatt, R., Goswami, A.: The big data system, components, tools, and technologies: a survey. Knowl. Inf. Syst. **60**(3), 1165–1245 (2019)
20. Ajah, I.A., Nweke, H.F.: Big data and business analytics: trends, platforms, success factors and applications. Big Data Cognit. Comput. **3**(2), 1–32 (2019)
21. Kazman, R., Bass, L.: Klein, M: The essential components of software architecture design and analysis. J. Syst. Software **79**(8), 1207–1216 (2006)
22. Lu, Q., Zhu, L., Xu, X., Xing, Z., Whittle, J.: Towards responsible AI in the era of generative AI: a reference architecture for designing foundation model based systems. IEEE Softw. **41**(6), 91–100 (2024)
23. Sena, B., Allian, A. P., Nakagawa, E. Y.: Characterizing big data software architectures: a systematic mapping study. In: Proceedings of the 11th Brazilian Symposium on Software Components, Architectures, and Reuse, pp. 1–10 (2017)
24. Ataei, P., Litchfield, A.: Big data reference architectures - a systematic literature review. In: ACIS 2020 Proceedings, vol. 30 (2020)
25. Avci, C., Tekinerdogan, B., Athanasiadis, I.N.: Software architectures for big data: a systematic literature review. Big Data Anal. **5**(1), 5 (2020)
26. Ataei, P., Litchfield, A.: The state of big data reference architectures: a systematic literature review. IEEE Access **10**, 113789–113807 (2022)
27. Brereton, P., Kitchenham, B.A., Budgen, D., Turner, M., Khalil, M.: Lessons from applying the systematic literature review process within the software engineering domain. J. Syst. Softw. **80**(4), 571–583 (2007)
28. Boell, S.K., Cecez-Kecmanovic, D.: On being 'systematic' in literature reviews in IS. Int. J. Inf. Technol. **30**(2), 161–173 (2015)
29. ISO/IEC/IEEE: ISO/IEC/IEEE 42010:2022 Software, systems and enterprise—Architecture description. International Organization for Standardization, Geneva (2022)
30. Hunnebeck, L. (2011). Service Design, London: The Stationary Office (TSO)
31. Reyes-Delgado, P.Y., Mora, M., Wang, F., Marx-Gomez, J.: AHP evaluation of rigorous and agile IT service design-building phases-workflows in data centers. J. Supercomput. **79**(16), 18089–18166 (2023)
32. Delen, D., Demirkan, H.: Data, information and analytics as services. Decis. Support. Syst. **55**, 359–363 (2013)
33. Wang, X., Yang, L., Liu, H., Deen, M.: A big data-as-a-service framework: state-of-the-art and perspectives. IEEE Trans. Big Data **4**(3), 325–340 (2017)

34. Angelov, S., Grefen, P., Greefhorst, D.: A framework for analysis and design of software reference architectures. Inf. Softw. Technol. **54**(4), 417–431 (2012)
35. Mora, M., Marx Gomez, J., Reyes-Delgado, P., Diaz, O.: An exploratory-descriptive review of main big data analytics reference architectures – an IT service management approach. In CONF-IRM 2022 Proceedings, vol. 6 (2022)
36. Mora, M., Gelman, O., Paradice, D., Cervantes, F.: The case for conceptual research in information systems. In: CONF-IRM 2008 Proceedings, vol. 52 (2008). http://aisel.aisnet.org/confirm2008/52
37. Brocke, J. V., Simons, A., Niehaves, B., Niehaves, B., Riemer, K., Plattfaut, R., Cleven, A.: Reconstructing the giant: on the importance of rigour in documenting the literature search process. In ECIS 2009 Proceedings, vol. 161 (2009)
38. Marz, N., Warren, J.: Big Data: Principles and Best Practices of Scalable Realtime Data Systems, 1st edn. Manning Publications Co., Greenwich (2015)
39. Stoica, I., Franklin, M., Jordan, M., Fox, A., Joseph, A., Mahoney, M., Katz, R., Patterson, D., Shenker, S.: The Berkeley Data Analysis System (BDAS): an open source platform for big data analytics. University of California, Berkeley (2017)
40. Chang, W. L., Boyd, D., Levin, O.: NIST Big Data Interoperability Framework: Volume 6, Reference Architecture v3. NIST, Gaithersburg, MD (2019)
41. Pääkkönen, P., Pakkala, D.: Reference architecture and classification of technologies, products and services for big data systems. Big Data Res. **2**(4), 166–186 (2015)
42. Cloud Standards Consumer Council: Cloud Customer Architecture for Big Data and Analytics V2.0. Cloud Standards Consumer Council (2017)
43. ISO/IEC: ISO/IEC 20547-3 Big data reference architecture - Part 3 Reference architecture. International Organization for Standardization (2020)
44. The Open Group IT4IT™ Reference Architecture, Version 2.1 (2017)
45. Clements, P., et al.: Documenting Software Architecture: Views and Beyond. Addison Wesley, Boston (2011)
46. The Open Group Architecture Framework (TOGAF). Version 10 (2024)

Chapter 2
Data Visualization in the Era of Data Science: A Review

Hector A. Duran-Limon, Arturo Chavoya, and Martha Hernández-Ochoa

Abstract The origin of data science dates back to the early foundations of statistics and probability in the early 60 s, which was then further developed with the rise of computing and database systems. However, it was not until the 2010s with the emergence of big data (i.e. large volumes of heterogeneous data) and the maturity of machine learning algorithms that data science gained enormous popularity. Data science is a discipline in charge of studying data with the aim of extracting knowledge from hidden patterns, descriptive analytics, and causal inference, among other sources of knowledge. Data science is an interdisciplinary field of study involving multiple areas, such as mathematics, statistics, and computer science. Data visualization, which is one of the key parts of data science, involves the process of creating a visual representation of data so that knowledge and insights are effectively communicated. There are different types of data visualizations, such as bar charts, histograms, line graphs, scatter plots, and heatmaps, among others. In addition, there are several tools for data visualization, such as Python libraries, R packages, and web-based tools (e.g. Tableau and Power BI). Nevertheless, there are a number of challenges in data visualization that need to be addressed, such as visualizing data with many variables, tools constraints (e.g. types of visualization and volume supported), and selecting the most suitable visualization type, among others. In this chapter, we review some of the most important types of visualization as well as the most widely used visualization tools. We also provide guides to select the most appropriate visualization type and visualization tool according to the volume and type of data. Finally, we point out some issues that need further research regarding data visualization in data science.

H. A. Duran-Limon (✉) · A. Chavoya
Information Systems, CUCEA, University of Guadalajara, Zapopan, Mexico
e-mail: hduran@cucea.udg.mx

A. Chavoya
e-mail: achavoya@cucea.udg.mx

M. Hernández-Ochoa
Knowledge Fundamentals, CUNORTE, University of Guadalajara, Colotlán, Mexico
e-mail: martha.ochoa@cunorte.udg.mx

M. Mora et al. (eds.), *Engineering and Management of Data Science, Analytics, and AI/ML Projects*, Intelligent Systems Reference Library 282,
https://doi.org/10.1007/978-3-032-06889-7_2

Keywords Data visualization · Data science · Visualization techniques · Visualization tools · Tableau · Power BI

2.1 Introduction

The roots of data science trace back to the early developments in statistics and probability during the 1960s [1]. Its evolution continued with advancements in computing and database technologies. However, data science did not experience widespread popularity until the 2010s, driven by the rise of big data—large, diverse datasets—and the maturation of machine learning algorithms [2]. It is worth mentioning some of the most important key milestones in the formation of data science. In 1957, Arthur Samuel coined the term "machine learning" and created one of the first successful self-learning programs. John Tukey, in 1947, coined the term "bit" at Bell Labs making a first intersection between statistics and computing. Also, in [3], Tukey advocated the integration of statistical methods with computational tools. In 1974, Peter Naur first used the term "data science" [4] and considered data processes for modeling reality. The formation of the International Association for Statistical Computing (IASC), in 1977, was a cornerstone of collaboration between statisticians and computing professionals. In 1986, Hinton together with Rumelhart and Williams authored a paper on applying the backpropagation algorithm to a multilayer neural network [5]. This represents a milestone as the algorithm allows the networks to learn internal representations of data. The NonSQL acronym was introduced by Carlo Strozzi in 1988 to refer to a database that did not use SQL. Later, the term was used to refer to non-relational databases. Nowadays, the NonSQL model uses a distributed database system that organizes large, unstructured data sets. The Knowledge Discovery in Databases organization held for the first time a data science workshop in 1989. The organization and conference were later renamed as the ACM SIGKDD Conference on Knowledge Discovery and Data Mining, which keeps running in 2025. Deep Blue, IBM's supercomputer, beat the world chess champion Kasparov in 1997. In 2002, the Data Science Journal was first published by the International Council for Science: Committee on Data for Science and Technology. Hadoop 0.1.0, an open source, non-relational database was released in 2006. Hadoop is able to handle large amounts of unstructured data. In 2011, the job post for data scientist had a huge demand. A year later, the Harvard Business Review declared the role of data scientist as the sexist job of the 21st century [2]. It is then recognized that the various social media applications generate huge amounts of data on a daily basis [6–8].

Data science is a field dedicated to examining data to uncover knowledge through hidden patterns, descriptive analytics, causal inference, and other insights [1]. The data can include both structured data and unstructured data. The former involves relational databases whereas the latter regards data in the form of text, images, and videos. As an interdisciplinary domain, data science integrates concepts from mathematics, statistics, and computer science [9]. There are different methodologies for data science such as KDD [10], CRISP-DM [11], and ASUM-DM [12], among others. In general terms, these methodologies involve processes to capture, maintain, explore, and analyze the data, which derive in a decision-making task. Hence, first, raw data is gathered from various sources. Then, the data is preprocessed whereby it is cleaned and transformed in the required format. More concretely, missing data is handled, duplicates are removed, inconsistencies are corrected, and values are normalized. The data is then explored to find the main characteristics of the data. The purpose of this step is to understand patterns, correlations, and outliers before modeling. This step is followed by an analysis of the data by using a model to generate reports and visualizations. In this step, such a model is created and evaluated for later use. A decision-making activity is then carried out based on the insights gained in the analysis. Therefore, the main role of data scientists involves collecting and cleaning data, developing predictive models, analyzing complex datasets to extract knowledge and insights from hidden patterns, and communicating such insights through visualization to the stakeholders.

Various types of data visualizations exist, including bar charts, histograms, line graphs, scatter plots, and heatmaps, among others [13]. Numerous tools are available for creating visualizations, such as Python libraries, R packages, and web-based platforms like Tableau and Power BI [14–16]. However, data visualization presents several challenges, including handling datasets with numerous variables, tool limitations (e.g., supported visualization types and data volume), and choosing the most appropriate visualization type for the data [14].

This chapter is organized as follows. We present some of the most important foundational concepts of data visualization. Next, we review some of the most important types of visualization as well as the most widely used visualization tools. Then, we provide a guide to select the most appropriate visualization type and visualization tool according to the volume and type of data. Finally, some concluding remarks are drawn.

2.2 Theoretical Foundations

In this section we present some principles of data visualization, along with cognitive and perceptual considerations, followed by a discussion on the Gestalt principles. These principles are depicted in Fig. 2.1.

2.2.1 Principles of Data Visualization

In data visualization, data is mapped to visual properties or attributes called *aesthetics*, which are represented by dots, lines, bars, and colors, among other properties [17]. The purpose of aesthetics is encoding multiple dimensions of data into a

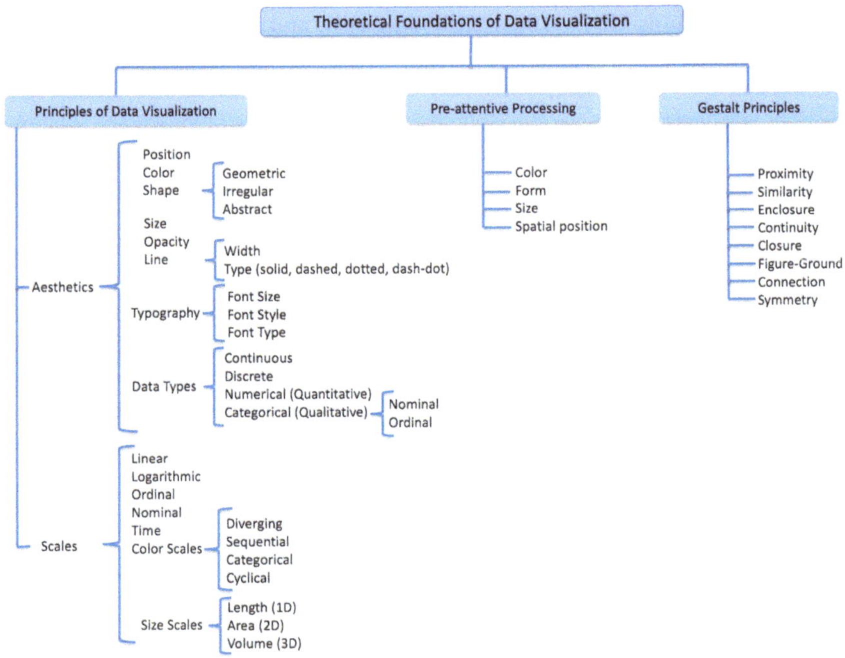

Fig. 2.1 Theoretical Foundations of Data Visualization

single data visualization. Aesthetics also aims to improve clarity and comprehension by employing visual cues to differentiate groups of data or identify patterns.

Below we describe some of the most common aesthetical graphical elements. A basic element is *position*, which determines the location of an element. The position in a 2D representation involves an x and y value. Nevertheless, more complex representations, such as 3D graphics, are possible in which a value for an extra dimension is defined. In addition, all graphical elements have *color*, *shape*, and *size*. Color plays an important role in provoking emotions and attracting attention. Color psychology tells us, for example, that red can be used for causing excitement whereas blue causes trust and calmness. Shapes are formed by lines and can be classified in three main categories. *Geometric shapes* such as triangles, squares, and circles can be used to define an order of how different graphic elements are related to each other. On the other hand, *irregular shapes* can be useful to highlight attention to specific areas in the visualization that can represent key findings or outliers. Irregular shapes are also helpful to represent real-world objects such as geographical areas. Lastly, *abstract shapes* can be used to represent concepts that do not have a real-world representation. Besides, size can be used to represent importance or hierarchy. Size is also helpful to represent quantity whereby a larger size represents a larger amount of something.

Another important aesthetic element is *opacity*, which is a characteristic that can be used to represent density or concentration of data points reflecting a higher or lower frequency. A higher opacity can also be used to highlight importance. Furthermore, *lines* can be used to visualize data. Lines can have different widths and there can be different types of lines. Examples of types of lines are solid, dashed, dotted, and dash-dot patterns. Different types of lines are helpful to differentiate from multiple lines in the same graphic whereas thicker lines can represent primary data series and thinner lines can display secondary data series. The *typography* is another important element of aesthetics that considers font size, font style, and font type. Larger font sizes emphasize more important information whereas smaller sizes are useful for secondary data. The font style can also be used to highlight importance, for instance when using a bold font. Besides, proper font type ensures legibility and clarity.

Aesthetics can represent two types of data, namely *continuous data* and *discrete data* [17]. Continuous data involves values for which infinite arbitrary intermediate values exist. For example, two-time values, let's say one second and two seconds, there can be infinite arbitrary intermediate values. In contrast, discrete data regards values that cannot have intermediate values. For instance, when the data refer to the number of people, there can be no intermediate values or fractional numbers. Data can also be either *numerical* or *categorical*. Numerical data, also called *quantitative data*, consist of measurable quantities and are represented by numbers. Numerical data can be continuous or discrete. It is possible to perform algorithm operations on numerical data. An example of numerical data is age or weather temperature. On the other side, categorical data, also called *qualitative data*, represent categories or groups. An example is product categories such as electronics, clothing, and furniture. Categorical data can be either *nominal* or *ordinal*. The former are categories that do not have an inherent order such a gender (e.g. male, female, no-binary). In contrast, ordinal data involves categories that have a meaningful order such as satisfaction level (e.g. poor, average, good, excellent). It is not possible to perform algorithm operations on categorical data. However, this type of data is usually converted to numeric values so that algorithms can be applied to it.

Scale is the system used to map data values to aesthetics such as position, shape, and size. Scales are important to translate raw data into a visualization that viewers can understand effectively. To avoid ambiguity, there must be a one-to-one mapping between data values and aesthetic values. For example, if we have data of the temperature of different geographical regions along different days, we can use three scales to do the mapping: we can map temperature values to the y axis, days to the x axis, and regions to colors. There are different types of scales. A *linear scale* maps values proportionally, that is, each unit increase in data results in a proportional increase in the visualization. In a *logarithmic scale*, each unit increase in data results in a logarithmic increase in the data visualization. A logarithm scale is useful when data involves a large range of values with exponential growth. In this case, a logarithm scale makes it easier to visualize differences in orders of magnitude that with a linear scale would be difficult to notice. In an *ordinal scale*, categorical data is mapped to specific positions or markers. The categories have an inherent order. For example, the service satisfaction level in a hotel can be represented by the following categories involving an order: bad, poor, good, and excellent. These categories can be represented by a bar in the x axis, while the number of occurrences of each category can be represented by height of the bar in the y axis. A *nominal scale* also maps categorical data to markers, but in this case the categories do not have an inherent order. An example of unordered categories are the product categories of a groceries store. A *time scale* maps temporal data to specific values. This type of scale is useful for representing time series data such as stock prices over days. In addition, in a *color scale*, data values are mapped to different colors. This type of scale can be helpful to represent complex datasets and highlight patterns. There are different types of color scales. A *diverging color scale* maps data that diverges from a neutral point to two opposite directions. The scale is commonly symmetric around the neutral point. This type of scale is useful to highlight the extremes of the data values with respect to the neutral point. A *sequential color scale* maps an ordered progression values to different light or brightness of a single color. For example, different color shades of blue can be used to represent the intensity of rainfall going from light blue to dark blue. *Categorical color scales* map categorical values to colors. *Cyclical color scales* represent data with periodic or repetitive patterns. In this type of scale, colors form a continuous loop of colors where the beginning and end meet seamlessly, i.e. there is no perceptual "start" or "end" of the scale. This type of scales is useful to represent periodic events such as seasonality, although they are also helpful to represent angles or directions (e.g. wind direction). Finally, *size scales* map values to different sizes of a particular shape. This type of scale is useful to represent either magnitude or intensity. Sizes can be expressed in three dimensions: one dimension i.e. length, b) two dimensions, i.e. area, and c) three dimensions i.e. volume. For example, a bubble chart can represent population by estate whereby a larger bubble represents a larger population.

2.2.2 Cognitive and Perceptual Considerations

It is important to use aesthetics to make relevant information intuitive and easy to digest. To this end, aesthetics must be adjusted to the characteristics of human perception. Next, we describe some of the most important characteristics of human perception. Pre-attentive processing [18, 19] refers to the unconscious pre-processing of information. The brain carries out this activity before conscious thought takes place. Pre-attentive activity takes 200–500 ms to transmit and process a visual stimulus, which is considerably faster than the time it takes conscious thought to process information. The pre-attentive process first gathers information from the environment and then filters out the most relevant information. Multiple stimuli can be processed in parallel without any conscious effort. Stimuli that stand out for their prominence are processed first and get more importance in the cognition. This prominence is influenced by certain characteristics such as color, form, size, and spatial position. For example, bright or contrasting colors can immediately attract attention. Also, unique forms and contrasting sizes can call attention first. In addition, the position of objects can guide attention.

2.2.3 *Gestalt Principles*

Gestalt principles involve a set of psychology rules or principles that explain how the human brain perceives and organizes data visualizations [20, 21]. These principles are relevant to understand how to organize and structure data visualizations for the viewer to effectively understand and interpret such data. The following are some of the key Gestalt principles applied in data visualization. *Proximity* establishes that elements that are placed next to each other are perceived as related. For example, a set of data points grouped closely indicates they are closely related. *Similarity* regards the fact that elements that look similar (e.g. shape, color) are perceived as belonging together. For example, elements belonging to the same category can be represented with the same color. *Enclosure* is the rule whereby elements that are within the same boundaries are conceived as being related. For instance, a set of elements placed within boxes indicates relationships. *Continuity* regards the principle in which continuous lines or patterns mean a consistent trend. For example, time series graphs with a continuous line are perceived as a trend pattern. The *closure* principle states that the human mind tends to fill in gaps in incomplete shapes to make sense of them. Minimalistic designs can, for example, be used for the viewers to infer the missing elements. The *figure-ground* rule establishes that elements are conceived either to be in the foreground or background. The former involves the main data whereas the latter is the context. For example, a dark line can indicate the main data whereas a light line may refer to contextual data. The *connection* principle states that elements that are connected by lines are conceived as related. For example, the lines used in a flowchart indicate relationships among the elements. Finally, the *symmetry* rule states that elements following a symmetry shape are considered as more balanced. For example, a graph of a normal distribution shows a balance in the right and left side of the shape.

2.3 Visualization Techniques

Some of the most used techniques for data visualization are presented next, with a brief description of the technique and the suggested use [17, 22, 23]. A more exhaustive list of techniques is presented in the *Guide to select visualization techniques* section. A brief discussion on the current state of 3D visualization techniques is presented as well.

2.3.1 *Charts and Graphs*

The selection of the appropriate graph or chart is of the utmost importance for an effective presentation of data, and this selection depends on various factors, such as data type, audience and intended message [23]. For instance, line charts are suitable to easily show time trends, whereas proportions are best presented using pie charts. More complex or multivariate data may require correspondingly more complex charts or graphs to show data relationships.

The most used categories of visualization techniques are presented in the following sections. For contrasting amounts for different datasets, *comparison charts* are among the most popular visualization techniques used, whereas *distribution charts* are typically used to display the spread of data for one or more categories. *Part-to-whole* and *hierarchical charts* allow to visualize relative proportions among datasets and *correlation charts* can show the possible connection between an independent variable and a dependent variable. For displaying datasets that are measured along a timeline, *data-over-time charts* can help show temporal trends, whereas spatial trends can be visualized using *geospatial* techniques. Also, when data measurements involve some kind of error, techniques that allow to visualize *uncertainty* can be utilized.

Examples of the visualization techniques described in the following sections are illustrated in Figs. 2.2 and 2.3. These examples were created using the Julius AI platform (https://julius.ai/) with data taken from the open-source repository found at https://vincentarelbundock.github.io/Rdatasets/datasets.html.

Comparison charts

They are used for comparing quantitative data across various categories, highlighting differences and similarities.

Bar and column charts

They allow comparisons to be made between discrete categories on a quantitative basis. Bar charts display the information horizontally, whereas column charts do it vertically. *When to use*: To visually differences among discrete categorical variables.

Dot chart

Dots from a dataset are placed in the appropriate position according to their values in the x and y axes. *When to use:* To replace bar charts when showing the bar height would be impractical or would obscure data features due to the range of values for the data points.

Grouped or clustered bar charts

They are used to compare mainly frequencies and counts across multiple groups or categories. *When to use*: To compare multiple groups or categories side by side.

Dumbbell plot

They display two points (the minimum and the maximum values) connected by a line for each category. *When to use*: To show distributions, variability and confidence intervals of the different categories.

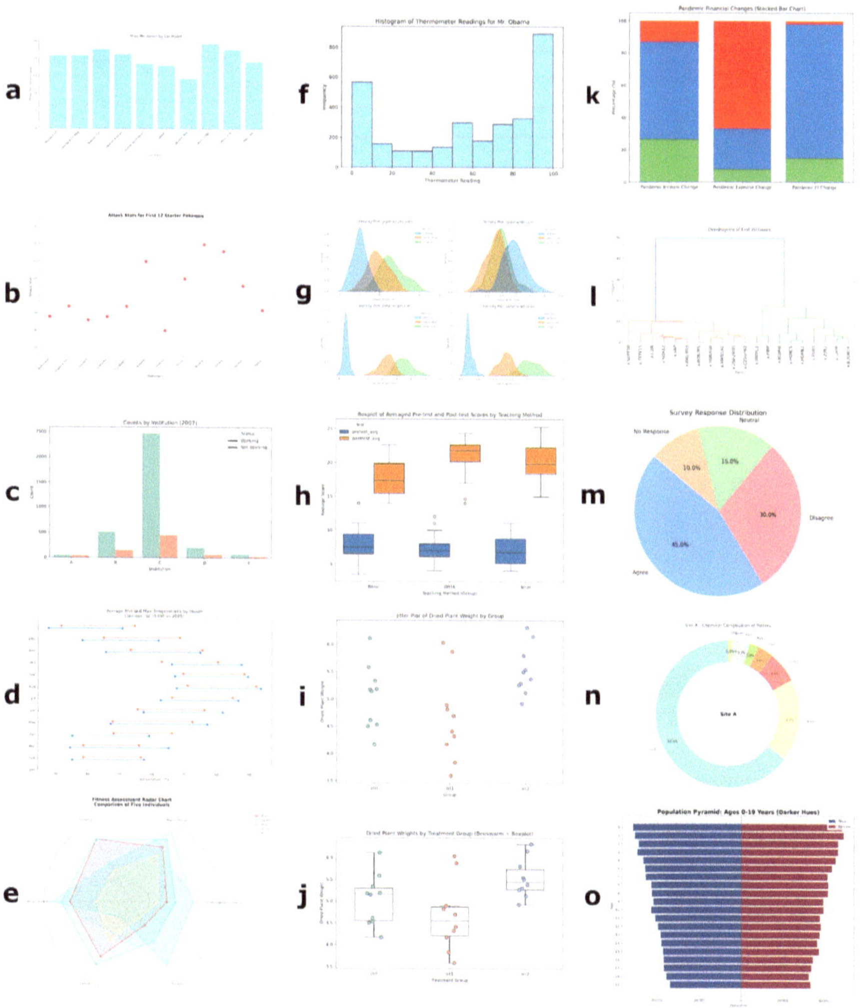

Fig. 2.2 Examples of comparison charts, distribution charts, and part-to-whole and hierarchical charts. **a** Column chart. **b** Dot chart. **c** Grouped bar chart. **d** Dumbbell plot. **e** Radar chart. **f** Histogram. **g** Density plots. **h** Box-and-whisker plot. **i** Jitter plot. **j** Beeswarm chart. **k** Stacked bars chart. **l** Dendogram. **m** Pie chart. **n** Donut chart. **o** Population pyramid

Radar chart

They overlap semi-transparent areas of different colors or hues whose vertices are connected across categories radially arranged starting from a common point. *When to use*: To compare multivariate data visually comparing the overlapping areas and vertices.

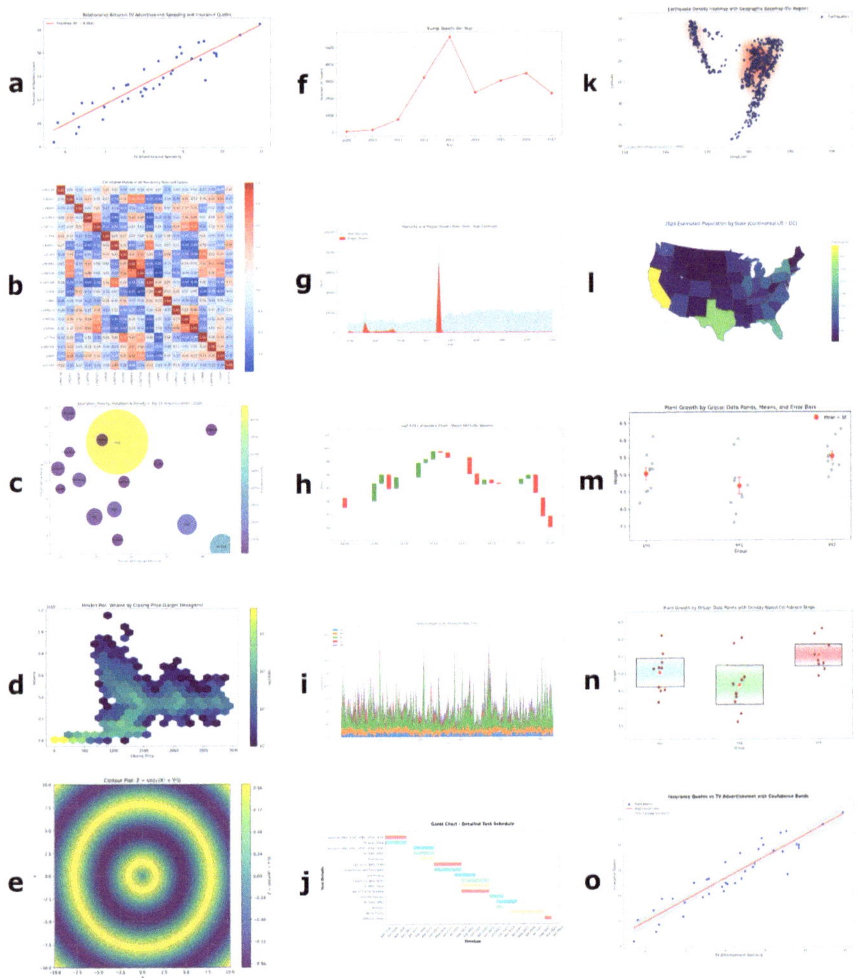

Fig. 2.3 Examples of correlation charts, data overtime charts, geospatial maps, and graphs for visualizing uncertainty. **a** Scatter plot. **b** Heatmaps and correlation matrices. **c** Bubble chart. **d** Hexagonal binning. **e** Contour plot. **f** Line chart. **g** Area chart. **h** Candlestick chart. **i** Stream graph. **j** Gantt chart. **k** Geographic heatmap. **l** Choropleth map. **m** Error bars. **n** Confidence strips. **o** Confidence bands

Distribution charts

They are used to show the spread of data over a number of categories or values. They can display frequency and range of data distributions. They can also show patterns and tendencies within the dataset.

Histogram

A dataset is divided into intervals or bins, where each of them is displayed as a bar whose height is proportional to the number of data points in that interval. *When to use:* To display the frequency distribution of a continuous variable.

Density plot

It plots the probability distribution of continuous variables, providing a smooth curve representing the distribution of data over a range. *When to use:* To show patterns and trends of the data's distribution.

Boxplot (box-and-whisker plot)

Normally displayed as rectangular boxes with vertical lines extending from the ends representing five data aspects in the following order: the minimum value, the first quartile (Q1), the median, the third quartile (Q3), and the maximum value. *When to use:* To display the distribution of datasets through a standardized method to highlight data features that represent the spread of data, the central tendency values, and the detection of outliers.

Jitter plot

Like a scatter plot but with an intentional random dispersion of points along one axis to avoid point clustering that could hinder a clear view of the representation of the data points. *When to use:* To show the density and distribution of data points that would normally overlap if presented in a scatter plot.

Beeswarm chart

Similar to a jitter plot displaying data points along one axis but only using a slight random dispersion when data points overlap. Although not necessary, beeswarm plots can include box plots or box-and-whisker plots to aid visualizing interquartile ranges. *When to use:* To show the density and distribution of multiple data points that would normally overlap due to their quantity.

Part-to-whole and hierarchical charts

Part-to-Whole charts show the contributions of individual portions to the whole, whereas Hierarchical charts reveal relationships between different levels of data represented in tree-like structures.

Stacked bars

Segments of each bar representing the different categories add up proportionally to represent the total amount. *When to use:* To show the composition of different categories in a dataset.

Dendogram

The hierarchical structure of clusters of data is displayed as a tree-like diagram. *When to use*: Typically used in biology, bioinformatics and machine learning to represent the relationships in a tree-like structure.

Pie chart

The total amount is represented by a circle with slices showing the relative contributions in proportion with the size of the slice. *When to use:* To show the relative proportions of different categories within a dataset.

Donut chart

Similar to pie charts, but using a donut shape, with the space left by the removed inner circle to be used to show some other information. *When to use:* To show the relative proportions of different categories within a dataset.

Population pyramid

Typically, they are bar charts arranged horizontally with age cohorts to the left or right, with one side representing females and the other males. *When to use:* To show the gender distribution of a population.

Correlation charts

They show how variables change in relation to other variables, indicating how close the relationship is and in what direction.

Scatter plot

The values from two variables are arranged as data points, each of which is represented as a point in a two-dimensional graph showing the relation of one variable against the other. *When to use:* To show the correlation between two variables using a set of data points.

Heatmaps and correlation matrices

They use color gradients representing variable values in a two-dimensional space. *When to use:* To visually identify patterns, relationships among variables and data anomalies.

Bubble chart

The position of each "bubble" (drawn as a circle) in the chart represents the intersection of two variables, whereas the radius of the circle represents another dimension of the data, with the possibility of presenting another dimension as a color gradient. *When to use:* To display multiple dimensions of data on a two-dimensional space.

Hexagonal binning

Each hexagon on a two-dimensional graph represents multiple dimensions of data given by its position and color gradient. *When to use:* For large and complex datasets comprising multiple continuous numerical variables.

Contour plots and surface plots

Contour plots represent three-dimensional data on a two-dimensional space plotting lines that connect points in common values on the third dimension. Alternatively, surface plots use planar surfaces with for instance a color gradient instead of lines for the common third dimensional variable. *When to use:* To show data distributions and densities on a two-dimensional field. They can also be used to show data as topographic maps.

Data over time charts

They are used to show data over time to highlight trends and patterns. They are useful in time series analysis and can be presented in a number of forms that depend on the type of data.

Line chart

One of the most used types of charts where the x-axis represents the time variable, and the y-axis shows the values of the dependent variable. *When to use:* To easily show trends and changes over time of one or more categories.

Area chart

The graph plots continuous values over time and the area below the plotted lines represents the magnitude of change over time. *When to use:* To represent quantitative continuous data and show changes of values over time.

Candlestick chart

The chart shows the high, the low, the opening and the closing prices of an asset over a particular time frame. *When to use:* To predict market trends.

Stream graph

It is a variation of the stacked area bar graph, but with the baseline freely moving up and down the graph according to the dependent variable value. *When to use:* To show changes of magnitude of different categories over time.

Gantt chart

It shows horizontal bars representing the start and the end dates of individual tasks or activities over time. *When to use:* Typically used in project management to visualize sequence, duration and dependence of tasks or activities over time.

Geospatial

Geospatial visualizations display data with geographic information, such as coordinates as longitude and latitude. They intend to relate spatial patterns with other information.

Geographic heatmap

This kind of map highlights the distribution of data in a certain geographic location, by using color gradients to represent density. *When to use*: To highlight trends, patterns and hot spots in spatial regions.

Choropleth map

It consists of a geographical map where areas are shaded representing values of a variable. *When to use:* To visually compare data distributions between areas of a particular spatial region.

Visualization of uncertainty

They are used to show data that can present some degree of error, which can be estimated as a range of values in which the actual data point may lie, as determined by probabilistic calculations.

Error bars

They are displayed as lines that extend horizontally or vertically to both sides of the datapoint to estimate the possible range of values for the datapoint. *When to use:* When a measure of probabilistic outcomes is needed along with the data points.

Confidence strips

The range of probabilistic outcomes for a data point is presented as a rectangle surrounding it where the density of its background represents the range of possible values. *When to use:* To replace error bars when a more detailed visualization of the possible range of values is needed.

Confidence bands

The range of possible values along a line graph are represented by the width of the band surrounding the line. *When to use:* To represent the probabilistic values for data points along a line graph.

2.3.2 3D Visualization

With the advent of big data and multivariate data, it might be necessary to display information in a spatial manner that can allow us to be immersed in the data and discover traits that otherwise would be difficult to discern. On the other hand, the falling costs and increased availability of head-mounted displays and other immersive technologies have raised the interest of researchers to present their data using these kinds of devices. 3D views can provide benefits over 2D displays; for instance, immersed views can provide context to the main data, 3D views can better model reality, and tangible visualizations can provide a physical interface through which a user can interact with the data. Furthermore, augmented reality can combine a tangible view with a virtual one, where the latter can provide detailed information in numerical or graphical ways [24].

Traditionally, three-dimensional structures have been represented as two-dimensional models, often suffering from potential loss of information due to dimensionality reduction, which can suggest the use of a 3D model instead [25]. However, 3D views can bring about problems not present in 2D visualizations. For instance, in a 3D view some data may be occluded by other data, or some data may be present in a position not viewed by the user, for instance behind them. Other problems can be the lack of context or scales of the viewed data. To overcome some of these problems, developers of 3D visualization tools should consider adding different views in addition to the 3D information. For instance, 3D data may be accompanied by color gradients representing changes in a range of values or presenting links in the data to present more detailed information of the particular point in space [24].

The field of 3D visualization as a general tool is still in its early development, and it is still not widely exploited; most current of 3D visualizations techniques are used in niche applications, such as medical imaging and surgical procedures [26, 27], chemical structure modeling of complex molecules [28], visualization of DNA sequences [29], forensic investigations [30, 31], and geological [32] and astronomical [33] modelling, among others.

However, as a general application tool, many problems to render useful information remain to be addressed in 3D visualization techniques, but proposals to solve some of these problems are currently underway, such as the use of a standard to store and render 3D representations [25].

2.4 Tools and Software

In this section we present some of the most important web-based platforms, followed by popular programming libraries for creating data visualizations.

2.4.1 Platforms

In this section we provide an overview of popular Web-based data visualization platforms, such as Tableau, Power BI, Google Charts, Infogram, Datawrapper, and Julius AI.

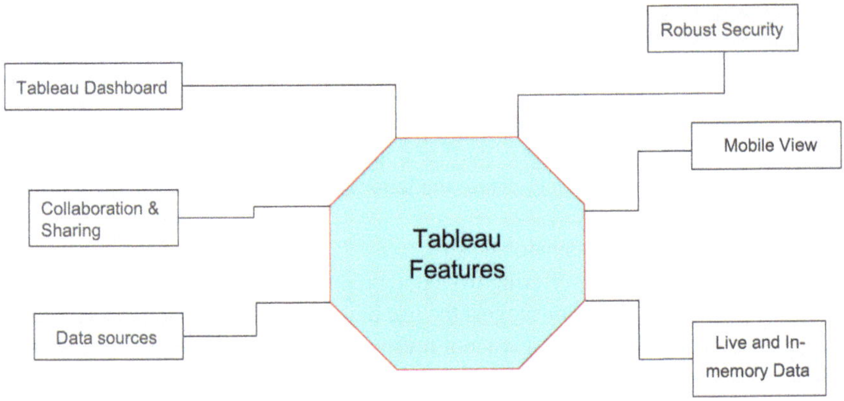

Fig. 2.4 Tableau features diagram

Tableau

Tableau was founded in 2003 as a result of a project at Stanford, whose main goal was to improve the flow of analysis and make data more accessible to people through visualization. Its mission remains to help people see and understand their data. Its major role is to design charts, create graphs and maps to perform data graphical analysis. Tableau is an intuitive tool that allows customers to approach analysis and visualization of data. This tool is focused on anyone (data scientist, professionals, business users, industries, engineering, among others), which uncovers data results faster with accessible machine learning, statistics, natural language, and smart data preparation [34]. Tableau is easy to learn by users and provides immediate feedback, and is also powered by AI (Artificial Intelligence). Its module of AI discovers relevant metrics, surfaces automated analytics in plain language, and proactively anticipates user questions. Some of the most important features of this tool are presented in Fig. 2.4 [34, 35].

Some of its advantages are real-time analysis, flexibility, intuitive interface, attractive data visualization, and a public version is available for free. Some drawbacks are its high cost for the desktop version [36], the public version only provides 1 GB of storage for online service, requires robust hardware, has compatibility issues (Desktop vs. Server), has slowed-down performance, is unsuitable for complex statistical analysis, and coding skills are required to work around R scripts for in-depth analysis [36].

Power BI

Power BI (Business Intelligence) is a reporting platform developed by Microsoft in 2013 [37]. The platform connects data stored in different Microsoft native software, such as Microsoft Common Data Service for Applications (CDS), Excel, SQL, Google Analytics, and MindChimp, which allows for easy organization to understand data reports. This tool is easy to use and collects information in a simple manner. Its main features are data connectivity, real-time data refresh, data modeling, customizability, data security, cloud-based, and simplicity to implement, among others [35]. Some advantages include ease of use, real-time data analysis, handling of large volumes of data, low cost, AI and machine learning functions, security, data control, and robust collaboration features. Some disadvantages are long-term costs, issues with non-standard data sources, the workbook size is limited to only to 250 MB, and is slower than Tableau [36].

Google charts

Google Charts is available on Google Play Store. It is a cloud-based data visualization tool whose main purpose is to manage and to create interactive charts that can be shared on the internet, unlike static charts. Typical customers are freelancers, small and mid-size businesses, and large enterprises, among others. Platforms supported are Web, Android and iPhone/iPad. Its most notable features are that there is a free version, charts have different formats, works with dynamic data, and renders via HTML5 and SVG [35, 38]. Some drawbacks are that it is dependent on Google services, advanced visualization is limited, large databases decrease its performance, and export options are limited.

Infogram

Infogram was created in 2012 by Uldis Leiterts, Raimonds Kaze, and Alise Semjonova [39], and now is part of Prezi. Visualizations can be exported to different file formats, such as PNG, JPG, GIF, PDF, and HTML, among others. Some main features are that it includes an API to allow additional data source to the database [35], has pre-design templates, enables animations that are easy to understand, has private or public publication, has different source integration data, is easy to explore, includes different types of graphics, and is ideal for embedding into websites or mobile applications. Some disadvantages are its lack of flexibility, high costs, real-time data limitation, and its free version is limited in terms of interactive features.

Datawrapper

Datawrapper is a tool that creates charts across different platforms in an easy way without assistance or additional training [40]. It does not require design skills. This tool helps sectors, such as government, global enterprises, and public or non-profit services. Its main features are that it has unlimited visualizations, and allows PNG, SVG, PDF format exports [40]. Some other advantages of this tool are that it is user-friendly, transparent, and does not require any coding or design skills. Some drawbacks are that it does not include advanced analysis tools or data manipulation, its exporting features are limited (e.g. no high-quality images), is not ideal for large datasets, and has limited integration with other tools.

Julius AI

Julius AI was developed in San Francisco, CA by Rahul Sonwalkar in 2022. This platform is an AI-powered data analyst and math helper [23]. This tool combines advanced technologies such as natural language processing (NLP), ChatGPT, machine learning (ML), and data visualization. In addition, Julius AI uses the next tools: Plotly, ggplot2, D3.js, Chart.js, and Matplotlib/Seaborn. This platform has different plans according to user needs including a free plan (15 requests per month). The main use cases in this platform are: finance, marketing, data science, scientific research, education, and business. The suitable tasks of Julius AI are: Python and R support, automated data exploration, data-driven decision-making, data analysis and visualization. Some key features are: charts and graph visualizations, data manipulation with simple prompts, linguistics analysis, data compatibility (CSV, PDF, Excel, Google Sheets, etc.), secure, reliable platform, and shared workspace, among others.

2.4.2 Programming Libraries

In this section we provide an overview of popular data visualization libraries that can be used with the Python, R and JavaScript programming languages.

Python

Python includes the following libraries: **(1) Matplotlib**—It generates charts and creates static, interactive and animated visualizations [41]; **(2) Seaborn**—It creates statistical data, allows for data exploration, has multiplot grids, and has automatic plotting [42]; **(3) Bokeh**—It has interactive charts, graphs, and web dashboards [43]; **(4) Plotly**—It has convenient functions and 3D visualization [44]; **(5) missingno**—It allows to view missing values and information, and displays correlations with heatmaps and demographics [45]; **(6) Folium**—It supports plotting interactive choropleth maps visualizations [46].

R

R is supported by the following libraries: **(1) ggplot2**—It supports univariate and multivariate visualizations [47]; **(2) ggvis**—It allows to build interactive graphics for empirical, exploratory and research data analysis [48]; **(3) Plotly R**—It allows to create interactive web-based graphs [49]; **(4) VIMGUI**—It allows to visualize univariate, bivariate, and multivariate graphs and plots [50]; **(5) VIM**—It helps to visualize missing values [51].

JavaScript

JavaScript can be used with the following libraries: **(1) D3.js**—It is fully integrated with web standards [52]; **(2) Recharts**—It is easy to use and creates interactive graphs, charts, and maps [53]; **(3) Victory**—It is intuitive, easy to use and creates basic charts and graphs [54]; **(4) react-vis**—It has a simple API for graph visualization and creates interactive charts [55], **(5) Chart.js**—It is easy to use, employs HTML5 elements, and is responsive and lightweight [56], **(6) ECharts**—It works with big datasets [57]; **(7) TradingVue.js**—It is a charting library for traders [58].

Although both Python and R are useful for data science and data analysis, the former has a stronger focus on artificial intelligence and web development, whereas the latter has a special strength on statistics. Python has a wider community support, libraries and frameworks; it is also easier to use and learn. However, Python is not suitable for embedded systems or mobile applications, as it is an interpreted language and is slower than C or C++. On the other hand, R has powerful libraries and a wider mathematical and statistical support. Nevertheless, R is less versatile and has a lower performance than Python. As for JavaScript, it is mainly used in web development and has better performance; however, it has some security issues and can be difficult to code in large projects.

2.4.3 *Comparison and Evaluation*

In this section we compare the tools and libraries based on the following criteria: open-source support, ease of use, flexibility, and performance for different tasks. These criteria are defined as follows:

Open source: Freely available for modification and redistribution, without having a corporate license.

User friendly: Easy to understand.

Flexibility: Easily adaptable for data visualization of different chart types; customizable aesthetics, such as color, visual elements, font, size, and labels, among others; ability to connect to multiple data sources; and real-time updates.

Performance: how effectively a data visualization can be created for the user´s experience, considering speed, memory requirements, accuracy, clarity, and accessibility.

Table 2.1 shows some features for the visualization tools and libraries selected from state-of-the-art literature, where a tick indicates that the tool fully meets the criterion, P indicates that the tool partially meets the criterion, and a cross means that the tool does not meet the criterion.

2.4.4 *Open-Source Versus Commercial Tools*

In this section we discuss the strengths and weaknesses of both open-source and commercial visualization tools.

Strengths

Open-Source tools: They are freely available to everyone, are suitable for developers that need to modify the library code according to specific needs, there is vast support in the open-source community, and are more suitable for research tasks. *Commercial tools:* They are user friendly, have customizable aesthetic flexibility, multiple predefined templates, personalized assistance, fast support, available technical manuals, high security level, can save time and effort and have better performance, and are more suitable for business.

Table 2.1 Comparison of visualization tools and libraries

Reference	Data visualization tool	Open source	User friendly	Flexibility	Performance
[22, 34, 35]	Tableau	X	X	✓	✓
[22, 35, 37]	Power BI	X	✓	X	P
[22, 35, 38]	Google Charts	X	✓	✓	✓
[22, 35, 39, 40]	Infogram	X	✓	X	✓
[22, 35, 40]	Datawrapper	X	P	P	P
[23]	Julius AI	X	✓	X	P
[22, 35, 41]	Matplotlib	✓	✓	✓	P
[22, 35, 42]	Seaborn	✓	✓	P	P
[22, 35, 43]	Bokeh	✓	X	✓	✓
[22, 35, 44]	Plotly	✓	✓	✓	P
[22, 35, 45]	missingno	✓	✓	P	✓
[22, 35, 46]	Folium	✓	✓	✓	P
[22, 35, 47]	ggplot2	✓	✓	✓	P
[22, 35, 48]	ggvis	✓	X	X	X
[22, 35, 49]	Plotly R	✓	✓	✓	X
[22, 35, 50]	VIMGUI	✓	P	X	✓
[22, 35, 51]	VIM	✓	P	✓	✓
[22, 35, 52]	D3.js	✓	✓	✓	P
[22, 35, 53]	Recharts	✓	✓	✓	P
[22, 35, 54]	Victory	✓	✓	✓	X
[22, 35, 55]	React-vis	✓	✓	✓	X
[22, 35, 56]	Chart.js	✓	✓	✓	X
[22, 35, 57]	ECharts	✓	✓	✓	X
[22, 35, 58]	TradingVue.js	✓	✓	✓	P

Weaknesses

Open-Source tools: They require programming abilities and technical skills and have security issues. *Commercial tools:* They typically can be expensive due to license, support, and subscription costs.

2.5 Guide to Select Techniques and Tools

In this section we present some recommendations in the form of guides to select the most appropriate visualization technique or visualization tool.

2.5.1 Guide to Select Visualization Techniques

The selection of the most appropriate chart or graph depends on the data, the audience, and the purpose of the visualization [23]. For instance, line charts are ideal to illustrate trends over time, whereas bar charts are more suitable for comparing different categories, and heatmaps can highlight correlations between variables. If simplicity for showcasing the data is the goal, bar charts are amongst the most popular types of graphical representation to compare categories, whether it is business data, research data or other types of data.

Table 2.2 presents a general guide to select the most appropriate type of visualization technique based on the nature of the data to be presented [17, 22, 23]. The first column corresponds to the type of visualization that it is desired, for instance, to visualize distributions or to display proportions; the second column indicates the use case that corresponds to the data to be visualized, for example, whether it is a single dataset or multiple datasets; finally, the third column allows to select among various visualization techniques that correspond to the selected visualization type and use case.

2.5.2 Guide to Select Visualization Tools

Nowadays there are many visualization tools in the software market, which makes selecting a tool a challenging job. Although each tool has its advantages and drawbacks, the selection of a tool depends on various factors, such as features, support, qualities, facilities, performance, flexibility, and costs. Some key factors on how tools can be selected depend on customer needs, organization goals, size of the datasets, features, data scope, user behavior data, audience, and intended message, among others. We recommend the following guide to select suitable tools (1) review the organization requirements, (2) identify the data scope, (3) identify costs, (4) determine customer needs, (5) review privacy requirements, (6) identify performance requirements, (7) identify flexibility requirements, and (8) determine the level of support required.

The Web-based data visualization platforms that we reviewed are able to create charts, maps, pie graphs, and help to carry out general graphical analyses. *Tableau* is preferred to handle large datasets and is recommended for advanced data visualizations. *Power BI* is suitable for seamless integration with the Microsoft Ecosystem. In addition, Power BI is more affordable, especially for small and medium-sized businesses. The *Google Charts* platform is utilized in Web pages, e-commerce, and education. *Infogram* gives solutions on issues of marketing, education, business, government, and social media, whereas *Datawrapper* is focused on non-profit and public services, media leaders, government, academy, and finance. *Julius AI* is particularly well-suited to situations where data analyst support is lacking. This is because

Table 2.2 Visualization techniques and their use cases

Visualization	Case	Technique
Amounts	For one set of categories	Bar charts (horizontal and vertical) Dot charts
	For two or more sets of categories	Grouped or clustered bars (horizontal and vertical) Stacked bars (horizontal and vertical) Heatmap Dumbbell plot Radar chart
Distributions	For basic distribution visualization	Histogram Density plot Cumulative density Quantile–Quantile plot
	For many distributions and/or for detecting shifts among the distributions	Boxplots Violins Strip charts Sina plots Stacked histograms Overlapping densities Ridgeline plot Jitter plot Beeswarm chart
Proportions	For analyzing one set of proportions	Pie chart Donut chart Bar chart (horizontal and vertical) Stacked bar chart (horizontal and vertical) Dendogram Population pyramid
	For multiple sets of proportions or changes in proportion across conditions	Multiple pie chart Grouped bar chart Multiple stacked bar chart Stacked densities
	For multiple grouping variables	Mosaic plot Treemap Parallel sets
x–y Relationships	For a small or moderate number of points	Scatterplot Bubble chart Paired scatterplot Slopegraph
	For large number of points	Contour plot Surface plot 2D Bins Hexagonal binning Correlogram Heatmaps Correlation matrices

(continued)

Table 2.2 (continued)

Visualization	Case	Technique
	When the x axis represents time	Line chart Area chart Candlestick chart Stream graph Gantt chart Connected scatterplot Smooth line graph
Geospatial data	For geospatial data in the form of a map	Map Choropleth map Cartogram Geographic heatmap
Uncertainty	For indicating the range of likely values for some estimate or measurement	Error bars (horizontal and vertical) 2D Error bar Graded error bars
	For a more detailed visualization of confidence ranges	Confidence strips Eyes Half-eyes Quantile dot plot
	For showing confidence as bands instead of bars in smooth line graphs	Confidence band Graded confidence band Fitted draws

the AI-powered data analysis and visualization of this tool can respond to natural language queries.

Regarding visualization libraries, *Matplotlib* produces quality plots, is a powerful collection of scientific tools, and is supported by astronomers at the Space Telescope Science Institute and NASA's Jet Propulsion Laboratory. In the case of *Seaborn*, the data is visualized by histograms, scatter plots, bar charts, pie charts and much more for scientific data and data analysts. *Bokeh* operates in a client/server model for the modern Web. *Plotly* is used in statistics, finance and scientific data, whereas *missingno* is employed for incomplete data patterns, and *Folium* is used for geographic and space data. *ggplot2* is utilized for economic, biological, statistics, and scientific data, and *ggvis* is like ggplot2, but for the web. *Plotly R* is used for exploratory data analysis with interactive graphics. *VIMGUI* is used for personalized simple interfaces, whereas *VIM* is preferred by data scientists given its programming flexibility, and its numerical and statistics support. *D3.js* is recommended for user interaction, such as brushing and linking. *Recharts* is used in React graphics, and *Victory* is focused on mobile applications, whereas *React-vis* is utilized for generating complex graphics. *Chart.js* is used in Web projects without React. In addition, *ECharts* focuses on high performance, interactive, and advanced graphics. Lastly, *Trading-Vue.js* is suitable for historic data costs to analyze financial markets in real time. In general terms, programming libraries are commonly used in research and academia and have enough flexibility to extend them to define new visualizations not supported by Web-based platforms, but they require programming skills.

2.6 Conclusions

Data science involves the study of data whose main goal is extracting knowledge from hidden patterns, descriptive analytics, and causal inference, among other sources of knowledge. The data may encompass both structured and unstructured formats. Information organized in relational databases is commonly regarded as structured data, while content such as text, images, and videos refers to unstructured data.

Data science is conformed by various study areas such as mathematics, statistics, and computer science. This convergence of multiple disciplines enables data scientists to address complex problems. There are different methodologies for data science that make it possible to drive decision-making across diverse industries. These methodologies involve processes to capture, maintain, explore, and analyze the data.

There are various forms of data visualizations, including bar charts, histograms, line graphs, scatter plots, and heatmaps, among others. Aesthetics, which are the attributes of data visualization, play an important role in the way a viewer digests the information contained in the visualization. It is important to take the cognitive and perceptual considerations along with the gestalt principles to make effective data visualizations. Such considerations and principles are not always completely intuitive; thus, we have defined a practical guide to select the most appropriate data visualization type according to the nature of the data that needs to be analyzed. For instance, those visualizations that involve trends over time are better visualized with line charts, whereas for discovering correlations between variables a visualization using a heatmap technique is more suitable.

In addition, there are numerous tools available for data visualization, including Web-based applications as well as Python libraries and R packages. We have reviewed the strengths and shortcomings of some of the most important tools used in data science. For example, among the Web-based applications, we found that Power BI is more affordable for small to medium-sized businesses, whereas Tableau is more suitable for handling complex and large datasets. Regarding programming libraries, Python libraries are suitable for problems involving data science, artificial intelligence, and Web development. Additionally, R packages are recommended for visualizing bioinformatics and statistics data.

Selecting the appropriate visualization type and the most suitable tool are current challenges in data visualization. We believe we have contributed with a practical guide to help address these issues. There are, however, other challenges that need further research, such as visualizing data with many variables. Although dimensionality reduction techniques can be used for visualizing data with many variables, there are a number of shortcomings involved in using these techniques. Firstly, a reduction can produce loss of important information. In addition, it can be computationally intensive, requiring significant processing power for large data sets. Finally, transformed features are commonly difficult to interpret.

References

1. Alazeb, A., Alshehri, M., Almakdi, S.: Review on data science and prediction. In: 2021 2nd International Conference on Computing and Data Science (CDS), pp. 548–555 (2021)
2. Davenport, T.H., Patil, D.J.: Data scientist: the sexiest job of the 21st century. Harv. Bus. Rev. **90** (2012)
3. Tukey, J.W.: The future of data analysis. In: Samuel, K., Johnson, N.L. (eds.) Breakthroughs in Statistics: Methodology and Distribution, pp. 408–452. Springer New York, New York, NY (1992)
4. Belzer, J.: Concise survey of computer methods. Peter Naur, 397p. Petrocelli Books, New York (1975). J. Am. Soc. Inf. Sci. **27**, 125–126 (1976). https://doi.org/10.1002/asi.4630270213
5. Rumelhart, D.E., Hinton, G.E., Williams, R.J.: Learning representations by back-propagating errors. Nature **323**, 533–536 (1986)
6. Rahman, Md.S., Reza, H.: A systematic review towards big data analytics in social media. Big Data Min. Anal. **5**, 228–244 (2022). https://doi.org/10.26599/BDMA.2022.9020009
7. Reinsel, D., Gantz, J., Rydning, J.: The digitalization of the world: from edge to core (2022)
8. Fisher, D., DeLine, R., Czerwinski, M., Drucker, S.: Interactions with big data analytics. Interactions **19**, 50–59 (2012). https://doi.org/10.1145/2168931.2168943
9. van der Aalst, W., Damiani, E.: Processes meet big data: connecting data science with process science. IEEE Trans. Serv. Comput. **8**, 810–819 (2015). https://doi.org/10.1109/TSC.2015.249 3732
10. Fayyad, U., Piatetsky-Shapiro, G., Smyth, P.: The KDD process for extracting useful knowledge from volumes of data. Commun. ACM **39**, 27–34 (1996). https://doi.org/10.1145/240455. 240464
11. Chapman, P.: CRISP-DM 1.0: step-by-step data mining guide (2000). https://api.semanticscho lar.org/CorpusID:59777418. Accessed 01 January–25 March 2025
12. Angée, S., Lozano-Argel, S.I., Munera, E.N.M., Ospina, J.D., Tabares-Betancur, M.S.: Towards an improved ASUM-DM process methodology for cross-disciplinary multi-organization big data & analytics projects. In: International Conference on Knowledge Management in Organizations (2018). Accessed 01 January–25 March 2025
13. Heer, J., Bostock, M., Ogievetsky, V.: A tour through the visualization zoo. Commun. ACM **53**, 59-67 (2010). https://doi.org/10.1145/1743546.1743567. Accessed 01 January–25 March 2025
14. Shakeel, H.M., Iram, S., Al-Aqrabi, H., Alsboui, T., Hill, R.: A Comprehensive State-of-the-Art Survey on Data Visualization Tools: Research Developments, Challenges and Future Domain Specific Visualization Framework. IEEE Access **10**, 96581–96601 (2022). https://doi.org/10. 1109/ACCESS.2022.3205115. Accessed 01 January–25 March 2025
15. Ali, S.M., Gupta, N., Nayak, G.K., Lenka, R.K.: Big data visualization: tools and challenges. In: 2016 2nd International Conference on Contemporary Computing and Informatics (IC3I), pp. 656–660 (2016). Accessed 01 January–25 March 2025
16. Mohammed, L.T., AlHabshy, A.A., ElDahshan, K.A.: Big data visualization: a survey. In: 2022 International Congress on Human-Computer Interaction, Optimization and Robotic Applications (HORA), pp. 1–12 (2022). Accessed 01 January–25 March 2025
17. Wilke, C.O.: Fundamentals of Data Visualization. O'Reilly Media, Inc. (2019). Accessed 01 January–25 March 2025
18. Atienza, M., Cantero, J.L., Escera, C.: Auditory information processing during human sleep as revealed by event-related brain potentials. Clin. Neurophysiol. **112**, 2031–2045 (2001). https:// doi.org/10.1016/S1388-2457(01)00650-2. Accessed 01 January–25 March 2025
19. der Heijden, A.H.C.: Perception for selection, selection for action, and action for perception. Vis. Cogn. **3**, 357, 361 (1996). Accessed 01 January–25 March 2025
20. Balan, S.: A systematic review of data visualization. J. Strateg. Innov. Sustain. **19** (2024). https://doi.org/10.33423/jsis.v19i3.7375. Accessed 01 January–25 March 2025

21. Guo, B., Huuki-Myers, L.A., Grant-Peters, M., Collado-Torres, L., Hicks, S.C.: escheR: unified multi-dimensional visualizations with Gestalt principles. Bioinf. Adv. **3**, vbad179 (2023). https://doi.org/10.1093/bioadv/vbad179. Accessed 01 January–25 March 2025
22. Shakeel, H.M., Iram, S., Al-Aqrabi, H., Alsboui, T., Hill, R.: A comprehensive state-of-the-art survey on data visualization tools: Research developments, challenges and future domain specific visualization framework. IEEE Access **10**, 96581–96601 (2022). Accessed 01 January–25 March 2025
23. Alysha Gullion: 29 best types of charts and graphs for data visualization. https://julius.ai/articles/types-of-charts-and-graphs. Accessed 01 January–25 March 2025
24. Roberts, J.C., Butcher, P.W.S., Ritsos, P.D.: One view is not enough: review of and encouragement for multiple and alternative representations in 3d and immersive visualisation. Computers **11** (2022). https://doi.org/10.3390/computers11020020. Accessed 01 January–25 March 2025
25. Mokos, D., Daniel, B.: Interactive 3D objects enhance scientific communication of structural data. ChemBioChem. n/a, e202500036 (2025). https://doi.org/10.1002/cbic.202500036. Accessed 01 January–25 March 2025
26. Bui, I., Bhattacharya, A., Wong, S.H., Singh, H.R., Agarwal, A.: Role of three-dimensional visualization modalities in medical education. Front Pediatr **9** (2021). https://doi.org/10.3389/fped.2021.760363. Accessed 01 January–25 March 2025
27. Venkatesan, M., Mohan, H., Ryan, J.R., Schürch, C.M., Nolan, G.P., Frakes, D.H., Coskun, A.F.: Virtual and augmented reality for biomedical applications. Cell. Rep. Med. **2** (2021). https://doi.org/10.1016/j.xcrm.2021.100348. Accessed 01 January–25 March 2025
28. Fombona-Pascual, A., Fombona, J., Vicente, R.: Augmented Reality, a Review of a Way to Represent and Manipulate 3D Chemical Structures. J. Chem. Inf. Model. 62, 1863–1872 (2022). https://doi.org/10.1021/acs.jcim.1c01255. Accessed 01 January–25 March 2025
29. Li, T., Li, M., Wu, Y., Li, Y.: Visualization methods for DNA sequences: a review and prospects. Biomolecules **14** (2024). https://doi.org/10.3390/biom14111447. Accessed 01 January–25 March 2025
30. Villa, C., Lynnerup, N., Jacobsen, C.: A virtual, 3D multimodal approach to victim and crime scene reconstruction. Diagnostics **13** (2023). https://doi.org/10.3390/diagnostics13172764. Accessed 01 January–25 March 2025
31. Ebert, L.C., Dobay, A., Franckenberg, S., Thali, M.J., Decker, S., Ford, J.: Image segmentation of post-mortem computed tomography data in forensic imaging: methods and applications. Forensic Imaging **28**, 200483 (2022). https://doi.org/10.1016/j.fri.2021.200483. Accessed 01 January–25 March 2025
32. Cao, X., Liu, Z., Hu, C., Song, X., Quaye, J.A., Lu, N.: Three-dimensional geological modelling in earth science research: an in-depth review and perspective analysis. Minerals **14** (2024). https://doi.org/10.3390/min14070686. Accessed 01 January–25 March 2025
33. Arcand, K.K., Price, S.R., Watzke, M.: Holding the cosmos in your hand: developing 3D modeling and printing pipelines for communications and research. Front. Earth. Sci. **8** (2020). https://doi.org/10.3389/feart.2020.590295. Accessed 01 January–25 March 2025
34. Tableau Team: Tableau (2025). https://www.tableau.com/en-gb. Accessed 01 January–25 March 2025
35. Mohammed, L.T., AlHabshy, A.A., ElDahshan, K.A.: Big data visualization: a survey. In: 2022 International Congress on Human-Computer Interaction, Optimization and Robotic Applications (HORA), pp. 1–12. IEEE (2022). Accessed 01 January–25 March 2025
36. Ali, S.M., Gupta, N., Nayak, G.K., Lenka, R.K.: Big data visualization: tools and challenges. In: 2016 2nd International Conference on Contemporary Computing and Informatics (IC3I), pp. 656–660. IEEE (2016). Accessed 01 January–25 March 2025
37. Microsoft Team: Power BI (2025). https://powerbi.microsoft.com/en-gb/. Accessed 01 January–25 March 2025
38. Google Team: emphGoogle chart (2025). https://developers.google.com/chart. Accessed 01 January–25 March 2025
39. Infogram Team: Infogram (2025). https://infogram.com/. Accessed 01 January–25 March 2025

40. Datawrapper Team: Datawrapper (2025). https://www.datawrapper.de/. Accessed 01 January–25 March 2025
41. Hunter, J., Dale, D., Firing, E., Droettboom, M., the Matplotlib Development Team: Matplotlib (2017). https://matplotlib.org/2.0.2/index.html. Accessed 01 January–25 March 2025
42. Waskom, M.: Seaborn (2021). https://seaborn.pydata.org/. Accessed 01 January–25 March 2025
43. NumFOCUS: Bokeh (2021). http://docs.bokeh.org/en/latest/. Accessed 01 January–25 March 2025
44. Johnson, A., Parmer, C., Sundquist, M.: Plotly (2025). https://plotly.com/. Accessed 01 January–25 March 2025
45. Bilogur, A.: Missingno (2025). https://github.com/ResidentMario/missingno. Accessed 01 January–25 March 2025
46. Python Software Foundation: Folium (2025). https://pypi.org/project/folium/0.1.5/. Accessed 01 January–25 March 2025
47. Wickham, H.: ggplot2 (2025). https://ggplot2.tidyverse.org/. Accessed 01 January–25 March 2025
48. Chang, W.: ggvis (2025). https://ggvis.rstudio.com/. Accessed 01 January–25 March 2025
49. Sievert, C.: Plotly R (2025). https://plotly.com/r/. Accessed 01 January–25 March 2025
50. Prantner, B., Schopfhauser, D., Templ, M., Alfons, A.: VIMGUI (2025). https://rdrr.io/cran/VIMGUI/. Accessed 01 January–25 March 2025
51. Templ, M.: Vim (2025). https://cran.rproject.org/web/packages/VIM/index.html. Accessed 01 January–25 March 2025
52. Bostock, M.: Data-Driven Documents (2021). https://d3js.org/. Accessed 01 January–25 March 2025
53. Bostock, M.: Recharts (2013). https://recharts.org/en-U.S./. Accessed 01 January–25 March 2025
54. FormidableLabs: Victory (2021). https://formidable.com/open-source/victory/. Accessed 01 January–25 March 2025
55. McNutt, A.: react-vis (2021). Accessed 01 January–25 March 2025
56. Timberg, E.: Chart.js (2021). https://www.chartjs.org/. Accessed 01 January–25 March 2025
57. The Apache Software Foundation: Apache ECharts (2017). https://echarts.apache.org/en/index.html. Accessed 01 January–25 March 2025
58. ChartMaster: trading-vue-js (2020). https://github.com/tvjsx/trading-vue-js. Accessed 01 January–25 March 2025

Chapter 3
Requirements for Using Federated Learning in Manufacturing Supply Chains

Jonas Kallisch and Jorge Marx Gómez

Abstract This chapter investigates Federated Learning (FL) as a data sovereignty preserving analytics option for manufacturing supply chains. Motivated by recent disruptions and the need for inter-company insight, it addresses four questions. At first which supply-chain areas gain the most from FL, second what best-practice implementations exist, third how FL can be integrated with current infrastructures, and fourth what product-, system- and chain-level requirements must be met. A systematic review of 21 peer-reviewed sources reveals five high-value application clusters—predictive maintenance, quality control, production optimization, collaborative planning and energy-management—where FL achieves near-centralized accuracy while safeguarding data sovereignty. The study distils prerequisites encompassing sensor granularity, edge computing, data standardization, governance and legal compliance, and discusses knowledge-distillation variants that further reduce trust barriers. Overall, FL unlocks cross-factory learning without exposing proprietary data, but its success depends on robust OT/IT integration, shared semantics and incentive-aligned consortia. The chapter closes with a research agenda and a process model to guide practitioners in deploying FL for resilient, competitive and sustainable manufacturing ecosystems.

Keywords Federated learning · Manufacturing supply chains · Data sovereignty · Production optimization · Supply chain collaboration

J. Kallisch (✉)
OFFIS, Oldenburg, Germany
e-mail: jonas.kallisch@offis.de

J. M. Gómez
Department of Informatics, Carl von Ossietzky University of Oldenburg, Oldenburg, Niedersachsen, Germany

M. Mora et al. (eds.), *Engineering and Management of Data Science, Analytics, and AI/ML Projects*, Intelligent Systems Reference Library 282,
https://doi.org/10.1007/978-3-032-06889-7_3

3.1 Introduction

Recent disruptions, such as global pandemics, geopolitical crises, and increasing demand volatility, have exposed some vulnerabilities in many manufacturing supply chains. Events like these highlight the necessity for deeper analytical insights across organizational boundaries.

Along with these external interruptions, studies have shown a growing demand for intercompany data analysis to boost supply chain efficiency [1], as supply chains increasingly require predictive capabilities and coordinated real-time decision-making to maintain productivity [2], manage complexity, and ensure resilience.

However, barriers related to data sovereignty, competition, and regulatory restrictions are limiting the usage of intercompany analytics. The heterogeneity of IT systems, varying data quality, and concerns regarding data ownership further complicate cross-enterprise collaboration, reducing the potential of conventional centralized data analysis approaches.

Particularly in Germany, where small and medium-sized companies form the core of industrial manufacturing structures and have built up process knowledge and intellectual property over long periods of time, there is a strong fear of losing control over such holdings. Nevertheless, these companies also see the need for cross-company data analysis in order to remain competitive internationally.

Federated machine learning (FL) offers a possible solution to this conundrum by enabling multiple organizations to collaboratively develop machine learning models without exchanging sensitive data [3–5]. By distributing model training across participants while keeping sensitive data local, FL addresses companies' concerns over data sovereignty. This solution is particularly relevant for manufacturing supply chains, where proprietary process data and intellectual property protection are paramount and legal challenges involving private data are not the main issue. While studies in banking, healthcare, and e-commerce confirm the practical advantages of FL, manufacturing companies remain cautious, primarily due to continued concerns about intellectual property, data sovereignty, and the potential risks of exposing strategic insights to competitors [3].

Nevertheless, early industry initiatives such as Catena-X and GAIA-X show the need for intercompany manufacturing collaboration. These initiatives focus on secure data spaces and standardized interoperability frameworks that enable companies to share data on a contract basis or with a mediator, like a data trust [4]. They align with the broader trend of digital ecosystems, where companies leverage shared data infrastructures while maintaining control over sensitive information [5]. However, in comparison with other sectors, manufacturing offers a few challenges to data ecosystems. These challenges include a much broader range of data types on the one hand, and on the other, the fact that there are often only small amounts of data that can be made available for analysis. This, combined with concerns about a loss of control, explains why there is no widespread use of machine learning in a cross-company context. FL could be a solution to address the mistrust between companies. However, a lack of examples currently limits its application in manufacturing supply chains.

To bridge this gap, further research is required to explore how FL can be effectively integrated into manufacturing supply chains, balancing analytical benefits with confidentiality requirements. This chapter investigates the specific potential of FL in intercompany manufacturing supply chains. It aims to provide an overview of the areas where FL could be used and what requirements individual companies, their products and their supply chains would need to meet to apply this technique.

3.2 Research Questions

This chapter addresses four research questions (RQ) regarding how and where FL can be used to improve the efficiency of manufacturing supply chains.

- RQ1: In which areas of a manufacturing supply chain does federated learning offer substantial benefits?

This question aims to understand the areas of a company where FL might be most able to improve the overall manufacturing supply chain's effectiveness. By identifying these, organizations can apply FL to the parts of their production system where it creates the most value.

- RQ2: Are there existing best practices or pilot projects demonstrating FL in manufacturing?

Investigating existing best practices and pilot implementations helps validate the practical applicability and feasibility of FL in real-world manufacturing scenarios.

- RQ3: What theoretical integration options exist for embedding FL within existing supply chain infrastructures?

Exploring theoretical frameworks and options for integrating FL into existing supply chain infrastructures is crucial for the implementation of intercompany analytics. Showing how FL can be combined with existing technologies reduces transition costs and promotes interoperability, enabling manufacturers to gain quick wins from the use of FL.

- RQ4: What product-, production system-, and supply chain-related requirements must be met to implement federated learning successfully?

Identifying necessary product, production system, and supply chain requirements clarifies the prerequisites for FL deployment. Defining these requirements upfront allows companies to prepare adequately, mitigating risks related to data incompatibilities and governance issues.

3.3 Research Method

To address these research questions, a meta-study approach based solely on a comprehensive literature analysis was used. This method is particularly suited for addressing the RQs, as it enables a systematic synthesis of existing knowledge and insights drawn from diverse academic and industry-related sources.

The literature analysis helps identify the fields where FL could be applied (RQ1) by examining studies and reports from industry and academia to extract common themes, success factors, and areas where FL has already demonstrated value. Regarding the investigation of existing best practices and pilot projects (RQ2), it provides a structured process for gathering case studies, empirical findings, and documented experiences from different manufacturing supply chains. Analyzing theoretical integration options (RQ3) through the literature review leverages existing frameworks and integration approaches, ensuring a broad understanding of interoperability, scalability, and compatibility issues across varied infrastructures. Lastly, the comprehensive analysis aids in extracting clearly defined requirements for successful FL implementation (RQ4), as it consolidates information from multiple authoritative sources, revealing essential technological, organizational, and operational prerequisites.

The search terms "Federated Learning," "Manufacturing Supply Chains," "Industry 4.0," "Data Privacy," "Data Sovereignty," and "Collaborative Learning" were selected based on their direct relevance to federated learning's role in enhancing analytical capabilities within industrial environments. While terms like "Federated Learning" and "Manufacturing Supply Chains" directly describe the primary scope, additional terms such as "Industry 4.0" and "Collaborative Learning" were included to ensure comprehensive coverage of interconnected topics and related technologies that are crucial for understanding practical implementation challenges and theoretical integration aspects.

Scopus, IEEE Xplore, and Google Scholar were chosen as primary databases due to their extensive collections of peer-reviewed journal and conference papers on technical, engineering, and management disciplines relevant to this study. ACM Digital Library and ScienceDirect were not included due to the initial scope and accessibility constraints during this preliminary research phase. However, the chosen databases provided adequate coverage to achieve comprehensive and representative results.

The search process identified 165 relevant publications. After screening based on defined inclusion criteria—such as direct relevance to FL in manufacturing contexts, peer-reviewed journal or conference status, and focus on practical implementation or theoretical integration—21 papers were selected for detailed analysis.

The final set of reviewed papers predominantly comprised peer-reviewed journal articles, supplemented by several conference papers, reflecting both established findings and ongoing exploratory research in the domain. The analysis identified three thematic clusters. The first includes papers and articles on the integration with Industry 4.0 and 5.0, addressing the alignment of FL technologies with advanced manufacturing paradigms. The second contains work on data privacy, sovereignty and security. These works explore methodologies for maintaining confidentiality and trust in decentralized data exchanges. The last cluster is summarized as work on challenges and future directions, highlighting barriers to FL adoption and suggesting potential solutions.

In the next section of this chapter, the authors will present the identified fields where FL could be used to improve intercompany data analysis in manufacturing supply chains.

3.4 Identified Applications for Federated Learning in Manufacturing Supply Chains

There are already a few examples of FL usage in the manufacturing sector across different operational fields. As FL allows multiple factories or devices to collaboratively train ML models without sharing raw data, there are naturally two kinds of scenarios where manufacturers could take advantage of this technique [3]. In the first scenario, companies take advantage of the data sovereignty provided by FL, which allows them to analyze intercompany data without sharing sensitive information [6, 7]. The second variant creates value by reducing the amount of data that needs to be shared [8, 9]. Many Industry 4.0 initiatives already leverage FL in areas like predictive maintenance and quality control to pool insights from distributed equipment while upholding data sovereignty. Emerging Industry 5.0 trends like human–robot collaboration and mass customization are also positioning FL as a mediator for analyzing intercompany data without exposing individual intellectual property. In the following sections, the chapter will examine the manufacturing applications of FL and show the benefits of this approach in the identified fields. Figure 3.1 gives an overview of the fields where FL could be applied in the manufacturing sector.

3.4.1 Different Kinds of Federated Learning

As described FL is a concept for analyzing datasets distributed over different devices connected with a central station [10]. It can be divided into horizontal FL, vertical FL and transfer Learning. The difference between vertical and horizontal FL is the selection of elements they share, while in transfer Learning a pre-trained model designed to perform one task, like detecting faults, is trained on another dataset to do something else, like identify fraud in an organization [11]. In the case of this chapter on supply chains, transferred learning is not relevant because generalized application scenarios could not be identified.

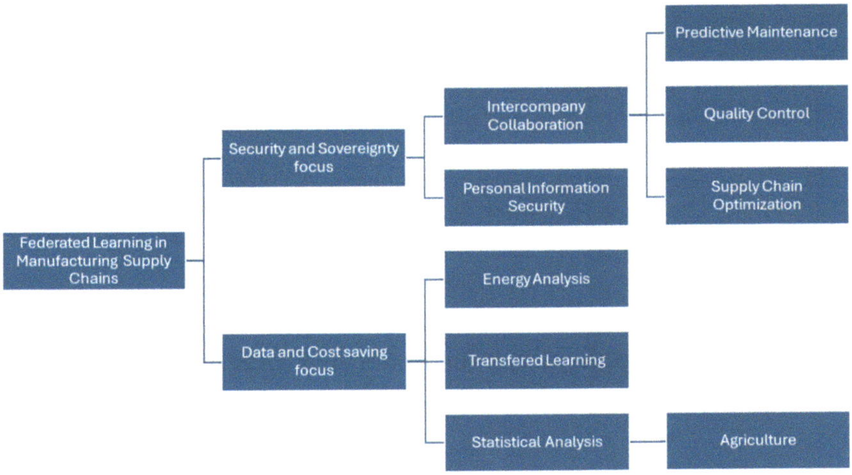

Fig. 3.1 Identified fields of application

As shown, horizontal FL shares features, e.g. temperature measuring or other data points, but not the samples—a concrete case of measurement, while vertical FL shares samples but not features.

The more common case of FL is the horizontal FL, as it is used in mobile devices to improve their ability to analyze user data without transferring it. A horizontal FL starts with an initial algorithm created on a data sample. In the second step, the model is decomposed into sub-models, matching the data elements of the different storages. The different data stores, e.g. smartphones, then train the model on their data. In the fourth and last step, the results of the individual models can be transferred to the central application to improve the model. The result is that the different data stores can improve their analytical models of their data without sharing. The limitation of the horizontal FL is that it requires similar data structures on different devices.

Vertical FL can be used in Supply Chains that do not share common data architectures between different companies. As shown in Fig. 3.2, the data stores in a vertical FL model do not share the same features—data structures—but the different data stores share the same samples. For example, two companies in the same city might not collect the same data but collect the data from the same customers. If these companies share a common interest, they could combine their data and use it to improve the quality of their prediction algorithms.

3.4.2 *Predictive Maintenance*

Predictive maintenance is one of the most mature FL applications within the manufacturing sector. Here, FL is used to train failure-prediction models across many

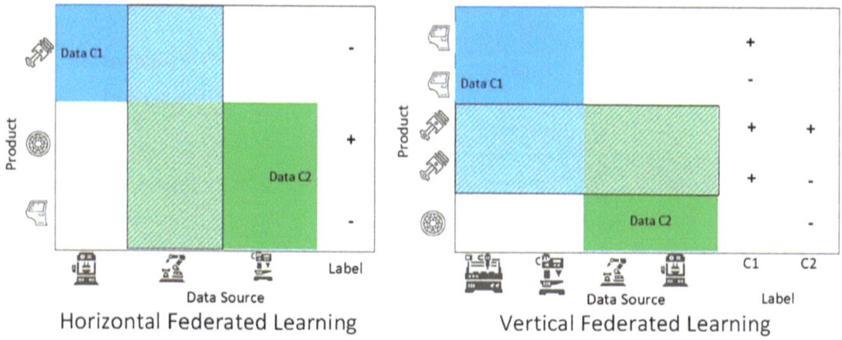

Fig. 3.2 Two types of federated learning

machines or even plants. For example, Siemens has applied FL to predict equipment failures across its factories. Each site trains a model on local sensor readings and shares only the model parameters to build a more accurate global model [12]. This collaborative approach improves fault detection by leveraging data from many machines, yet no raw maintenance logs or proprietary usage data leave the local facility. An automotive pilot by Stellantis showed that FL on welding robot data from multiple factories yielded a deeper understanding of equipment operations and helped anticipate robot failures, all without centralizing sensitive process data [6].

Predictive maintenance models that use FL tend to be more accurate and robust than models trained at a single site, since they combine a wider range of operating conditions and failures from different systems and companies. This is shown in Fig. 3.3, as all manufacturers who maintain the same kind of robotic system can now use data and event information from 8 Systems instead of the smaller number they own. Manufacturers report reduced unplanned downtime and longer equipment lifespans as a result of applying FL [7]. The main value of FL in this field is that it enables this accuracy without sacrificing data sovereignty or incurring huge data transfer costs—only model updates and not the raw sensor data are exchanged, preserving confidentiality and bandwidth [8]. FL lets even rival firms or an OEM and its customers build shared predictive models without exposing their individual machine usage patterns, which could reveal trade secrets or production volumes. In fact, equipment manufacturers are beginning to deploy FL "analytic agents" to client machines to learn failure indicators that would otherwise be inaccessible due to confidentiality. The OEM can then offer improved maintenance services back to all clients based on the aggregated model. Compared to a traditional centralized approach, with a central aggregation of the raw data of each partner, FL provides better predictive insights for all participants and better planning capabilities for machine manufacturers, with minimal data sharing risk, while eliminating the costs of large data centers needed to store data from different sources [9].

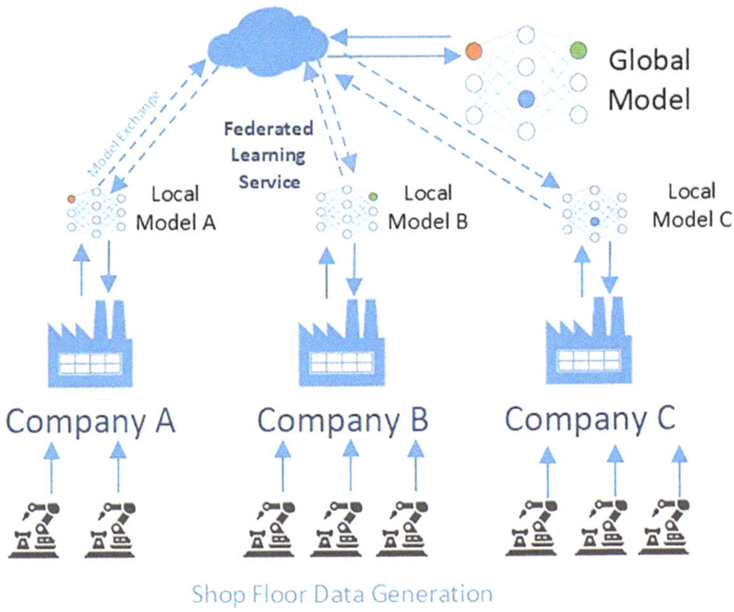

Fig. 3.3 Horizontal federated learning in predictive maintenance

3.4.3 Quality Control and Defect Detection

Quality control in manufacturing companies is often limited by the amount of labeled data for the different types of failures that can occur [9]. FL offers a solution by aggregating quality inspection information across multiple production lines or companies, even rivals. In practice, each factory trains a local model and shares the learned parameters to form a stronger global model. Federated quality control has been demonstrated to improve defect detection rates because the global model is exposed to a much broader set of defect examples and process variations than any single site could provide [13]. This means subtle quality issues caught in one plant help the model identify issues in other participating plants. Therefore, FL addresses the data scarcity problem for training high-accuracy inspection models. Moreover, by keeping all images and quality data on the shop floor, manufacturers maintain strict control over sensitive information, which is especially important if multiple companies collaborate on improving quality standards [14].

Real-world examples underscore these benefits. Siemens recently used FL to connect quality inspection data from two of its electronics factories to improve an automated optical inspection (AOI) system for printed circuit boards [13]. Before FL, they had a centralized AI (XGBoost) that only worked well on specific PCB variants and required retraining for each site. After introducing FL, the AOI's false-positive rate dropped significantly and the federated model generalized across different PCB types and locations [13]. This reduced costly manual re-inspections and scrap, and

Siemens is now rolling the FL-driven quality model out to additional factories without needing to export any local production data [13]. In another study on additive manufacturing quality, researchers applied FL to predict the dimensional accuracy of 3D-printed parts across up to 45 distributed printers [15]. The federated model could predict part dimensions within 10 μm of a fully centralized model's accuracy, and it outperformed models trained at individual sites by up to 96%—a dramatic improvement in part quality prediction achieved through privacy-preserving collaboration [14]. These cases show that FL can yield nearly the same quality control performance as traditional data-centralized training, while each factory keeps its process data secured.

Another practical benefit of using FL in quality control is that, by combining the quality data of different manufacturers, it helps enforce more uniform quality standards across production sites [13]. Plants with limited defect samples benefit from others that have seen those defects, leading to earlier detection of anomalies and fewer defective parts reaching customers. FL also cuts down the enormous data transfer that a centralized vision inspection system would require—high-resolution images and sensor streams stay at the edge, and only lightweight model parameters move, greatly reducing bandwidth and storage demands [13]. This reduces costs in the fields of data communication and infrastructure. Keeping sensitive data local further reduces the costs of infrastructure, as there is no need for the security of the FL platform to also secure the intellectual property of the plants and companies.

3.4.4 Production Process Optimization

Manufacturing companies are eager to identify the best operating practices across their network of plants. However, direct sharing of raw process data, especially between independent companies, remains limited. One reason is that these data elements contain sensitive information, and sharing this information can cause serious disadvantages for a company. In addition, the exchange can be problematic due to differences in IT systems, data formats, or simple internal reluctance to expose local inefficiencies to others [16]. As FL provides a way to analyze data without directly sharing it, process optimization can be undertaken while respecting boundaries and circumventing the shortcomings of standardized data formats. In an intra-company setting, each factory can use its own historical process data to train a local model. A federated aggregation then produces a global model that essentially captures what works best across all plants, without requiring any plant to share its raw process data. This model can reveal insights like which combinations of parameters lead to optimal outcomes, helping all participants improve their production systems based on a global view. This kind of analysis could even be carried out across an entire supply chain to improve its overall production efficiency, rather than simply adjusting to local optima [17].

The benefits are clear in terms of efficiency and knowledge transfer. FL enables a "comprehensive understanding of process strengths and areas for improvement" across the companies involved [13]. For example, if one factory has mastered a more efficient welding technique, the federated model will learn the correlation between that factory's settings and its high throughput. Other factories can then apply that insight via the improved model recommendations—without ever directly seeing the first factory's data. This preserves each site's autonomy and protects trade secrets [7].

3.4.5 Supply Chain and Intercompany Collaboration

As described earlier, besides its usage within the production system of a single manufacturing company, FL could be used across an entire manufacturing supply chain and its intercompany manufacturing processes to improve the overall production system. As in most supply chains, each company has its own valuable data, like demand forecasts, cost structures or machine data from their shop floor, that cannot be shared, but could improve overall planning if combined. Reasons for this lack of direct data sharing are manifold, from fear of competition to confidentiality issues, mainly concerning the sharing of personal information. FL offers a mechanism for collaborative analytics in such cases. For instance, suppose that multiple suppliers and a manufacturer jointly train a demand forecasting model, with each firm using its own sales and supply data locally. The FL framework merges these models into a united model and sends updates to each participant. The resulting intercompany model can predict demand or supply disruptions more accurately for all, without any participant seeing the raw data of another [7]. This approach has been proposed for inventory optimization and logistics as well, where retailers, distributors, and producers train shared models to streamline deliveries and reduce bullwhip effects while keeping proprietary operations data private [18].

The advantages of FL in supply chain scenarios also include improved accuracy. Thanks to the diverse data of the different companies involved, a federated supply chain model can better anticipate issues in the production system and relocate resources to the areas where they will create the most value [14]. These usages of supply chain-wide information can increase the efficiency of the overall production system instead of creating only local optima on the individual shop floors. Therefore, FL could enable supply chains to run more smoothly, reduce the required storage capacities and increase the competitiveness of the entire supply chain.

Beyond demand and logistics, FL can facilitate intercompany production process optimization in value chains. Consider a scenario where one company's output is another's input, which is very common in manufacturing supply chains. Tuning the end-to-end process for efficiency is challenging if each firm only optimizes its own production system. With FL, companies can train a joint model on data from each step, yielding insights that account for interdependencies between steps [19]. For example, a parts supplier and an assembly manufacturer might collaboratively train a model to minimize final product defects using the supplier's process data and the manufacturer's quality data. FL allows both companies to identify how variations

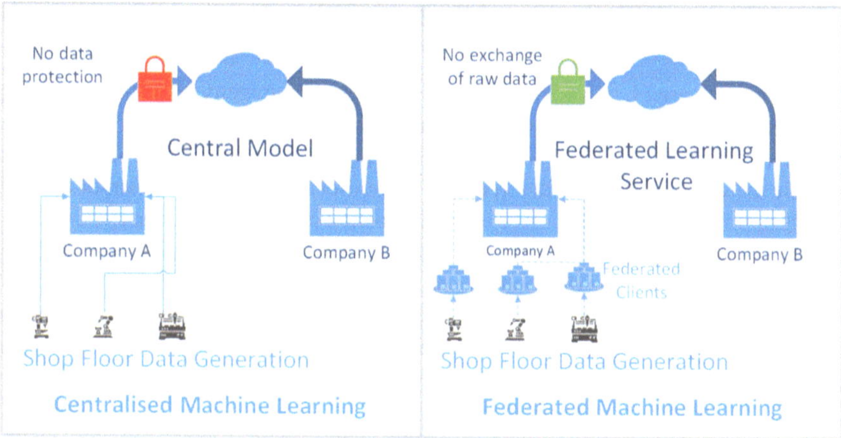

Fig. 3.4 Centralized versus federated learning in supply chains

in the production system affect the assembly quality on the other end of the supply chain. This enables them to adjust the system based on the information from the whole chain, without risking the loss of control over their individual intellectual property and data sovereignty. The difference in the protection of intellectual property between traditional, aggregated machine learning and the approach of FL is shown in Fig. 3.4. With holistic optimization FL can eliminate the inefficiencies that isolated optimization creates, leading to better performance for the entire chain [13].

In summary, FL unlocks a capacity for intercompany cooperation that was previously limited by data sovereignty and competition, making it a powerful alternative to centralized data exchanges or data trusts in manufacturing ecosystems.

While governance structures and trust agreements are central to enabling FL across supply chains, practical challenges remain regarding whether companies are willing to share even aggregated model data with an orchestrator. In many real-world scenarios, concerns about revealing sensitive patterns or strategic insights persist. To address this, alternative approaches such as model distillation can be considered. Model distillation involves training a smaller global model based on the knowledge distilled from local models, without directly aggregating or transferring raw parameters or data. Recent initiatives, such as DeepSeek, have demonstrated how knowledge can be extracted from multiple source models without exposing underlying proprietary information. Integrating such techniques into FL frameworks may increase acceptance among industrial partners by further minimizing data exposure risks.

3.4.6 *Energy Efficiency and Sustainability*

Another area where FL can be beneficial is the energy management of manufacturing companies. Modern factories are commonly instrumented with energy meters and IoT sensors to track electricity and other energy resource usage. By applying FL, multiple factories can train a shared model to predict energy consumption or identify savings opportunities without disclosing their actual consumption data. For example, a group of energy-intensive plants could use federated learning to develop a model that forecasts peak power demand or recommends load balancing strategies based on production schedules. Each plant's data stays on-site, but the aggregated model helps all participants operate more efficiently [20]. This kind of collaboration can be orchestrated by a third party that provides the FL platform or by an industry consortium focused on sustainability.

The foremost benefit is improved accuracy in energy forecasting and optimization. As the federated model includes a wider number of operating patterns, it can therefore predict energy usage more accurately [7]. This enables better planning and can cut costs. One practical example is using federated learning for predictive maintenance of HVAC or boiler systems across buildings—the same idea applies in factories to anticipate when equipment is using more energy than normal. FL also accelerates the learning of best energy-saving practices. This means that if one facility discovers an optimal way to, say, schedule machine start-ups to avoid a demand surge, the federated model will capture that behavior and help others adopt it, all without any direct communication of what each facility did. Additionally, from a sustainability reporting perspective, FL allows collective progress without data pooling. Companies are under pressure to reduce their carbon footprints, and FL can help by facilitating joint models that locate efficiency gains while each company keeps its data confidential, which is important if energy data could reveal production secrets. Compared to centralized energy analytics, FL avoids the need to centralize potentially gigabytes of sensor data per day from each site, reducing infrastructure burdens and security risks. It also ensures data sovereignty: multinational manufacturers can keep their energy data within local servers, complying with local energy reporting rules, and still contribute to a global optimization effort. Ultimately, federated learning for energy management aligns well with industry sustainability goals by pooling intelligence, not data, to cut resource waste [21].

3.5 Identified Requirements for the Use of Federated Learning in Manufacturing Supply Chains

To answer the question of what prerequisites are necessary for the use of federated learning in industrial supply chains, an analysis of FL's known concepts and applications was carried out. The results of this analysis identified three clusters of requirements that establish applicability in different scenarios. These are product-specific, production system-specific and supply chain-specific requirements, and are described below.

3.5.1 Product-Specific Requirements

Implementing FL in a manufacturing supply chain demands careful consideration of the data generated at the product and machine levels. This section of the chapter will investigate the product-related requirements to use FL in a manufacturing system.

3.5.1.1 High-Resolution Data and Granularity

FL models perform best when each participant—for example, a machine or production line—can provide sufficiently granular data about the product or process of manufacturing. Fine-grained sensor readings enable the detection of subtle patterns, whereas coarse, aggregated data may obscure important variability. In practice, varying data granularity across factories is a known challenge, so aligning standards for an appropriate level of detail is crucial [22]. For example, if one plant only logs batch-level summaries but another logs each unit's metrics, the global model may not generalize well. Thus, each product or part should ideally be tracked with consistent detail to feed the FL analysis. The literature showed that the granularity of data provided by the different companies involved had a major influence on the results of the application.

Similarly, as the companies involved in an analysis will have different infrastructures and data schemes, a common data structure is needed to use FL methods. Companies also have to consider the different data inputs of the participants to avoid the phenomenon of "unbalanced clients." This term describes a situation where one member of the group enacting an FL application generates more information value than the others. As a result, the FL model may undervalue the data this company provides and therefore not deliver the best possible results.

3.5.1.2 Sensor and IoT Infrastructure

A robust sensor network is needed on the shop floor to capture the relevant data for model training. Machinery must be equipped with sensors to monitor operating conditions and product quality in real time [23]. These IoT devices provide the local data that each FL client will use for training. Ensuring that legacy equipment is retrofitted with modern sensors or that newer machines' built-in sensors are accessible is a key requirement. A real-world example can be found in the area of predictive maintenance. In this use case, FL has been employed to build failure-prediction models across a network of factories, with each machine's sensor data being processed locally to predict failures and only model updates being shared. Without a capable sensor infrastructure generating timely, relevant data, FL's potential for improving product outcomes, like quality and reliability, cannot be realized.

3.5.1.3 Data Quality and Labeling

The quality of local data must be high, since noisy or erroneous data can degrade the global model. In manufacturing environments, data may be incomplete [24]. Reasons for this include sensor dropouts or maintenance downtime. In addition, data might be inconsistent due to sensor drift or calibration differences [25]. Each participant should have processes to clean and validate their data before it is used in training. In cases of supervised learning, consistent labeling of outcomes is needed across sites, which may require common defect codes or quality criteria. Notably, studies have found that if one partner's data quality is substantially lower than others', it can negatively impact the federated model more than if all partners have uniformly lower but consistent data quality [26]. This implies a requirement for minimum data quality standards among collaborators or mechanisms to weight or filter out low-quality data during training. Techniques like outlier removal, normalization, and synchronization of data timestamps across sensors are practical measures to enhance data quality. Ultimately, garbage in yields garbage out—therefore, ensuring high-quality, well-labeled data at the product level is foundational for FL success.

3.5.1.4 Data Quantity

While the benefits of federated learning in manufacturing are compelling, a key practical challenge that emerged during the analysis relates to data availability, specifically the quantity of labeled data required to train effective FL models. A recurrent issue in real-world industrial settings is the limited amount of data—particularly the scarcity of labeled error or defect cases [26].

Most manufacturers generate vast amounts of process data, but much of it remains underutilized, especially when it comes to identifying and labeling edge cases or production anomalies [27]. In many companies, product data is only retained or flagged when a part meets quality standards, while non-conforming products are either discarded or not systematically tracked. This practice limits the amount and diversity of datasets needed to train federated machine learning models.

Moreover, smaller manufacturing facilities or those with low-volume, high-complexity products face the challenge of insufficient amounts of data in general [28]. In such environments, even over extended periods, the number of fault cases may be too low to create a robust model locally. FL, while promising, still requires a minimum critical amount of data on each participating node to ensure that the local models contribute meaningfully to the global model.

These challenges are exacerbated by the imbalance in class distribution—most industrial datasets are dominated by normal operation data, with defective or abnormal cases forming a small minority. This imbalance can result in biased models that fail to recognize rare but critical events. In a federated context, where models are trained locally before being aggregated, this imbalance can dilute the contribution of rare event data even further if only a subset of participants has access to it.

3.5.2 Production System Requirements

At the production system or factory level, implementing federated learning demands significant infrastructure and integration efforts. The following requirements must be met within each production facility or node in the supply chain to take advantage of FL's potential. However, as the individual requirements are also necessary for other future business models and technologies, they do not represent any significant restrictions. Such business models include, for example, manufacturing as a service, virtual company networks or interactive product development. All of these use cases are based on becoming the best supplier rather than the most cost-effective.

3.5.2.1 Edge Computing Capability

Each factory should possess enough local computing resources to perform training and inference tasks on-site. FL shifts the computational burden to the "edge," so participants need sufficient processing power and memory to train models on their portion of the data. In practice, this might mean deploying an industrial edge platform or upgrading existing shop-floor computers to handle ML workloads. The benefit is reduced latency and bandwidth usage, since only model parameters (not raw data) are communicated [7]. This kind of architecture is shown in Fig. 3.5. However, it introduces a trade-off, as the company must invest in distributed computational resources rather than relying on a single powerful central server.

That said, recent research highlights that incorporating edge devices for FL in IoT can minimize energy and communication costs, leading to a more sustainable computing environment [29]. In addition, distributed application systems offer manufacturing companies other advantages, as they make applications such as the asset administration shell easier to integrate. This allows new types of order management to be applied, more potential for optimization to be identified and internal communication and transparency to be increased. Summed up, a capable edge computing infrastructure is a prerequisite to train models locally and participate effectively in FL, but it also delivers the foundation for new opportunities in factory control and communication.

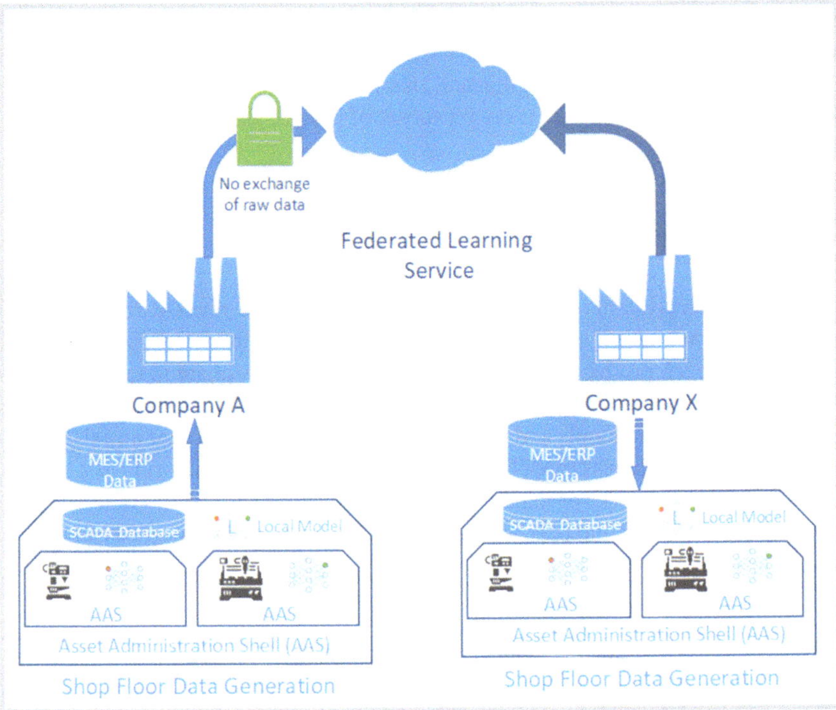

Fig. 3.5 Aggregation of models over different layers of the production system

3.5.2.2 OT/IT Integration and Data Pipeline

A seamless integration between operational technology (OT) on the factory floor and information technology (IT) systems is essential. Data from sensors and machines (OT) must be collected, aggregated, and made available to the FL training process in near real-time. This often requires developing data pipelines that connect shop-floor devices, such as PLCs, SCADA systems or IoT gateways, with edge servers or cloud networks. For example, in Siemens' implementation of FL for quality control, each factory built a multi-layer architecture connecting physical devices to a local edge computer, then further up to a cloud-based FL server [13]. At the field level, connectors pulled PLC data into an AI inference server and data collector module; at the factory level, an FL agent and model manager interfaced with the central platform [13]. This illustrates the need for middleware, APIs, or message brokers, most likely OPC-UA, that bridge shop-floor data to the FL software. Additionally, integration with existing manufacturing execution systems (MES) or other databases can provide the historical data needed for initial model training. The challenge is that many factories have heterogeneous systems and legacy equipment. These systems must be integrated into a broader model to take full advantage of FL.

3.5.2.3 Local Data Preprocessing and Conversion

Before training in the FL model, raw sensor data typically needs to be filtered, transformed, and stored locally. This means each site should implement preprocessing steps such as denoising sensor signals and adding missing values. Doing this close to the source both reduces noise in the model and ensures that only relevant information is used, minimizing unnecessary communication. Managing "unclean" factory data is a known hurdle, and if data is not cleaned and standardized, an FL model may struggle to converge [22]. For instance, if one machine reports temperatures in Celsius and another in Fahrenheit, a conversion from one to the other must be performed so that the model sees a consistent measurement. Similarly, aligning timestamps and production IDs might be necessary when products pass through multiple machines. Factories may use local data lakes or databases to store batches of processed data for training iterations, and these have to be discussed with all partners to develop an intercompany data history strategy. Researchers have suggested integrating data cleaning and efficient data selection as part of the FL pipeline to address this issue [22]. In short, each participant needs a data preprocessing pipeline, a conversion into a common format and adequate local storage to feed high-quality inputs into the federated model.

3.5.2.4 Reliable Network and Connectivity

Federated training is an iterative, communication-intensive process—models are exchanged potentially hundreds or thousands of times during an FL training round and will be again in further iterations. Therefore, a reliable network connection is required between the participants and the FL aggregator. Bandwidth limitations and latency can significantly slow down the exchange process and the reconfiguration of the FL model [7]. Ideally, participants should have redundant network paths or offline caching to handle intermittent connections. In an industrial campus, upgrading to Wi-Fi 6 or private 5G networks may be considered to guarantee low-latency, high-throughput data exchange. Communication efficiency techniques, like data compression, mitigate the bandwidth burden, but they do not eliminate the need for a stable connection.

3.5.3 Supply Chain Requirements

When extending federated learning across multiple organizations in a supply chain, additional requirements come into play at the ecosystem or network level. These relate to how different companies collaborate, share, and govern the FL process. A common understanding of the process is necessary for efficient collaboration. Therefore, besides the product and production system requirements, some supply chain requirements also need to be met in order to successfully implement FL in manufacturing supply chains. This section will give insight into those identified requirements.

3.5.3.1 Data Interoperability and Standardization

To collaboratively train a model, all participants need to feed in data that the model can understand in a unified way. This requires agreeing on common data definitions, including what features are used by the model and how they are measured. If one supplier logs temperature data in degrees and another with another scale, the model input must be standardized [30]. Ensuring data interoperability often involves establishing industry-wide data schemas or using existing standards for manufacturing data, like OPC-UA [31]. In supply chain settings, interoperability also extends to IT systems. For instance, the FL platform might need to interface with each firm's data sources, which could be ERP systems, quality management databases, or IoT platforms [32]. Using common interchange formats or middleware to convert different formats can facilitate this. Semantic alignment is needed as well—for example, all factories involved should use the same definition of a defect or a cycle time, if those are model features. The participants as a group might agree and publish templates for data to be used in collaborative analytics. In the semiconductor industry, it was noted that data sharing between different process tools must be seamless and data provenance is critical for digital twins and analytics [33]. This equally applies to FL, as each partner must deliver data in a form that can be integrated and trusted. This might entail additional metadata accompanying model updates, describing how the data was collected or its quality metrics, so that the aggregator can interpret contributions correctly.

3.5.3.2 Governance Structures and Trust Agreements

Unlike intraorganizational FL, where one organization has full control, multi-company FL requires a governance framework. Participants must establish who orchestrates the training—ideally a neutral party or the principal manufacturer in the chain—and how the global model and updates are managed. A culture of trust and open collaboration is needed, since each party is being asked to contribute insights without seeing other participants' raw data [34]. Creating trust may involve legal agreements and technical assurances in addition to the FL system. One practical approach is to use a data alliance model, where all members join a federation with clearly defined benefits and responsibilities [7]. Prior research suggests that developing a business model for a data-sharing alliance in parallel with the technical FL mechanism is crucial [35]. This means stakeholders should agree on questions like: Who owns the resulting model? Can a participant opt out at any time? How are improvements measured and credited? It also extends to decision-making. For example, if the model needs to be updated or retrained, do all partners need to consent? In large networks, hierarchical governance may be needed to answer to this question.

Some emerging ecosystems might provide a template for such governance options, as Europe's International Data Spaces (IDS) architecture does [36]. In the IDS, only certified participants can join a certain data space, and they can attach usage policies to their data/models [37]. The IDS architecture ensures that data sovereignty is respected by all parties, which aligns well with FL's assurance that raw data stays local. Each participant might run an IDS connector that interfaces with the FL platform, enforcing the agreed policies. In essence, FL governance at the supply chain level must handle both technical coordination and relationship management. Therefore, establishing a consortium or at least a contractual framework is a prerequisite to multi-organization FL.

3.5.3.3 Governance Structures and Alternative Model Training Approaches

FL inherently minimizes the need for direct data sharing, addressing many concerns around data sovereignty. However, the governance structures surrounding model training and aggregation remain critical. The traditional approach of having a central orchestrator aggregating local models requires robust governance frameworks to build trust among participants. Even though, some companies may still have reservations regarding indirect information leakage or the power dynamics inherent in an orchestrator-led model.

To mitigate these concerns, alternative methods such as model distillation or knowledge transfer techniques can be employed [38]. In such approaches, companies train their models locally and subsequently distill the knowledge into a compact representation or lightweight models, rather than directly aggregating model parameters. Techniques like federated distillation or ensemble approaches allow participants to collaboratively enhance model performance without exchanging raw data or even raw model parameters explicitly [39]. These non-invasive procedures substantially lower barriers to adoption by ensuring that no raw or granular data ever leaves the local infrastructure, thereby further preserving data confidentiality and company autonomy.

A notable example of this is the model distillation approach used in systems like DeepSeek, where models are trained based on knowledge distilled from multiple local models without needing direct data exposure [40]. Such techniques could also serve manufacturing supply chains well, offering a feasible governance model that promotes trust and aligns incentives across diverse participants.

3.5.3.4 Collaboration Models and Incentives

Along with governance, a practical collaboration model should be defined. There are different ways that companies might take advantage of FL's potential. One model is a centralized federation where a lead enterprise, for example an OEM, provides the FL server and invites suppliers to join, which gives the lead more control and requires others to trust that party. Another model is a third-party platform, which can be provided as a data trust by an third company who hosts the FL aggregator in a neutral cloud environment. In this variant, all participants are clients to the platform. This approach has been used in projects like MUSKETEER, an EU initiative, and can increase trust by not giving any single manufacturing company privileged access [37]. The MUSKETEER platform combined federated learning with privacy-preserving tech and was tested in a manufacturing use case for automotive suppliers [41]. In that scenario, multiple factories collaboratively trained a welding quality prediction model, with the platform handling encryption and attack mitigation to ensure fair play [41]. Such collaboration models illustrate the importance of aligning incentives, as each participant should see values, like improved model accuracy or better operational insights, that outweigh any remaining risks and additional costs. A scenario from the MUSKETEER Platform is shown in Fig. 3.6, where the manufacturers test classification is combined analyzed with the machine data.

Sometimes incentives might be contractual: when, for example, a group of companies jointly fund the FL infrastructure or give discounts on services for contributing data. Another approach is reciprocity, where all parties get access to a better model than they could train alone, which is often incentive enough if the use case is well chosen, like collaborative predictive maintenance reducing downtime for all.

In summary, designing a collaboration model involves not just technical network topology, but also business governance that encourages continued participation and contribution. Transparency about the model's performance, contributions and security can help maintain trust in the long run.

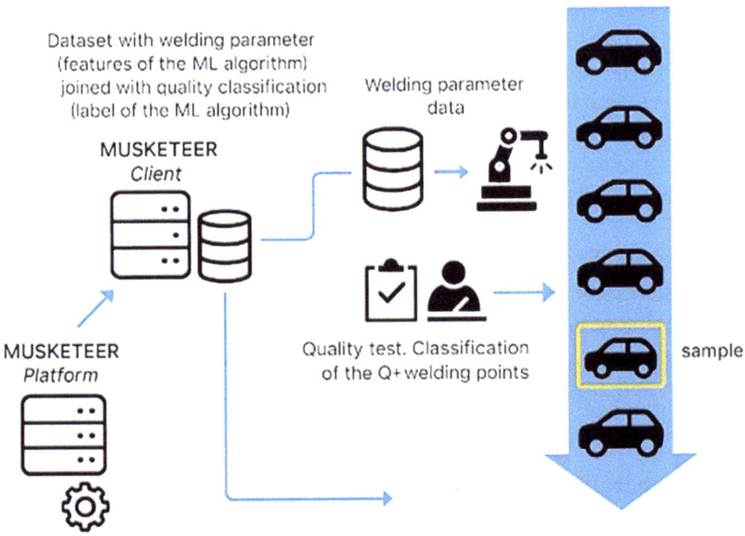

Fig. 3.6 Usage scenarios for the MUSKETEER platform (based on [42])

3.5.3.5 Data Sovereignty, Compliance and Legal Considerations

Sovereignty is the central motivator for FL, but when dealing with supply chain data, there are specific compliance issues to address. First, if any of the data can be linked to individuals, like worker performance data, or sensor data that might include information about operators, regulations like GDPR or CCPA come into play. FL, by design, keeps personal data local, aligning with the principle of data minimization, as it was originally conceived to tackle privacy issues with distributed data [7]. However, companies still must ensure that no sensitive data or business-critical information can be reconstructed from the shared model. Techniques discussed under the production system requirements—for example, if differential privacy is needed—become even more critical in a multi-company context, because here there is not only a risk of external attackers: rather, even a collaborating company might be considered a potential adversary. Thus, the FL protocol should be strengthened against information leakage. Research indicates that additional mechanisms are needed to prevent advanced attacks like model inversion or membership inference between participants [43]. Using privacy-preserving technologies and a framework that prevents indirect information leakage is paramount.

In essence, while FL significantly mitigates risk by never pooling raw data, it does not eliminate accountability—organizations must still demonstrate compliance and due diligence in how the federated training is conducted. This includes obtaining any necessary consents, carrying out privacy impact assessments, and aligning the FL initiative with the overall data governance strategy of the supply chain.

3.5.3.6 Mapping Data Relationships Across the Supply Chain

Mapping data relationships is essential to successfully implement FL across manufacturing supply chains. This mapping clarifies where which data is generated, how it relates to shared workflows. In addition the mapping should clarify what insights can be collaboratively derived without compromising data sovereignty. Especially for small and medium-sized enterprises (SME), such transparency is a prerequisite for trust and participation in any intercompany analytics initiative.

Existing supply chain relationships naturally generate interlinked data—for instance, a part produced by Supplier A is later assembled and quality-tested by Manufacturer B [44]. Traditionally, these data streams remain siloed, as firms are hesitant to share sensitive information that could reveal intellectual property, competitive advantages, or inefficiencies [45].

Drawing from prior research and prototype development in secure information exchange systems, a concept has been introduced in which data relationships are not based on centralized sharing but instead on linking information via process support platforms. These platforms use workflow connectors to trace how data originating from one partner (e.g., A1 from Supplier A) influences outcomes at another partner (e.g., B2 from Manufacturer B), without requiring the raw data to leave the premises [46].

In practice, each participant holds its own data (A1 or B1) and derives insights (A2 or B2). Only the insights, which can be filtered or anonymized, are exchanged through the platform based on bilateral or multilateral service agreements [44]. This mapping model allows each company to retain full control over their intellectual property while still benefiting from system-wide improvements.

In this setting, federated learning can be layered on top of these process-based relationships. For example, if anomalies are detected in the quality control data at OEM B, the system can trigger a federated training round in which local models at Supplier A and OEM B collaboratively improve failure prediction models—all without sharing underlying process logs or sensitive production parameters.

This form of value-oriented relationship mapping not only makes intercompany FL technically feasible but also builds the foundation for trust-based data ecosystems. The platform concept developed at the "Zukunftslabor Produktion" (Future Lab Manufacturing) supports this architecture by linking partners through containers for communication, services, and analytics. Each containerized application acts independently but can be orchestrated for shared tasks when predefined triggers or contracts allow it. The architecture used by the Futurelab Manufacturing is shown in Fig. 3.7.

By mapping supply chain data relationships around shared processes and product flow, rather than shared databases, the FL approach aligns naturally with the decentralized and heterogeneous reality of manufacturing SCs. It also supports selective insight sharing, anomaly tracing, and even root cause diagnostics across company borders—all while upholding the principle of data sovereignty.

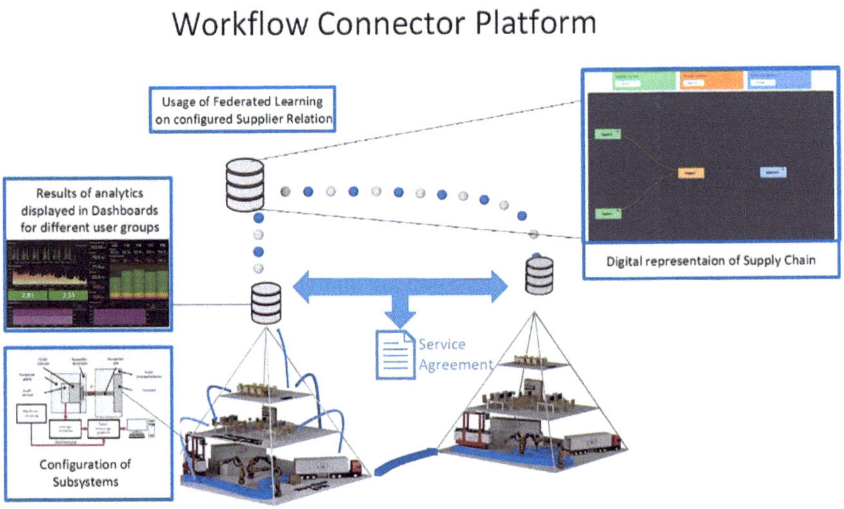

Fig. 3.7 Approach of using federated learning at the futurelab manufacturing [47]

3.6 Discussion and Conclusion

This chapter explored the application potential, best practices, integration possibilities, and essential requirements for adopting federated learning (FL) into manufacturing supply chains. The systematic literature analysis underscored significant opportunities for FL to positively impact manufacturing, notably in predictive maintenance, quality control, production optimization, supply chain collaboration, and sustainability.

FL has shown considerable promise in manufacturing, as it addresses data sovereignty and intellectual property concerns inherent in those environments. The examined pilot projects and industry examples, such as those from Siemens and Stellantis, validate FL's feasibility and effectiveness in real-world scenarios. They illustrate how companies can jointly improve operational efficiency, defect detection accuracy, energy management, and forecasting capabilities without compromising data confidentiality or incurring prohibitive data-sharing risks.

Despite these promising outcomes, the successful integration of FL into existing manufacturing infrastructures presents several challenges. Primary among these are ensuring data quality, standardization, and interoperability across heterogenic manufacturing systems and data infrastructures. Overcoming these issues necessitates robust sensor networks, reliable edge computing infrastructure, seamless OT/IT integration, rigorous data preprocessing, and trust between the companies interacting with each other.

Additional challenges in expanding FL's application across entire supply chains include creating governance frameworks, defining clear collaborative models, aligning incentives, and ensuring strict regulatory compliance. Practical solutions identified through initiatives like GAIA-X and Catena-X offer a template for constructing trust-based collaborative ecosystems that effectively manage these complexities.

Ultimately, for FL to realize its full potential in manufacturing supply chains, organizations must proactively address technological, organizational, and legal prerequisites. This requires an interdisciplinary approach, bringing together expertise from IT, operations, legal, and supply chain management to build resilient, privacy-preserving data ecosystems.

In conclusion, while FL represents an innovative technique that could help enable enhanced predictive analytics and collaborative optimization in manufacturing, its successful implementation demands careful preparation, comprehensive infrastructure investment, and strategic governance frameworks. Further empirical research is necessary to validate current theoretical insights in practical settings. Additionally, the development of a comprehensive process model for applying FL in manufacturing contexts would greatly assist organizations in systematically addressing implementation challenges and effectively leveraging FL's advantages for enhanced supply chain resilience and competitiveness.

References

1. Hermes Germany GmbH: Transparenz in der supply chain. Supply X Barometer **13**, 1–24 (2019). https://supplyx.info/wp-content/uploads/2024/09/supplyx-barometer-13-supply-chain-transparenz.pdf
2. Sahay, B.S., Ranjan, J.: Real time business intelligence in supply chain analytics. Inf. Manag. Comput. Secur. **16**, 28–48 (2008). https://doi.org/10.1108/09685220810862733
3. Li, L., Fan, Y., Tse, M., et al.: A review of applications in federated learning. Comput. Ind. Eng. **149**, 106854 (2020). https://doi.org/10.1016/j.cie.2020.106854
4. Stachon, M., Möller, F., Guggenberger, T., et al.: Understanding data trusts. In: Proceedings of 31st European Conference on Information Systems. Kristiansand, pp. 1–17 (2023)
5. Hussain, F.K.: Digital ecosystems and technologies. In: Proceedings of 2nd IEEE International Conference on Digital Ecosystems & Technologies (DEST). IEEE, Phitsanulok, pp. 26–29 (2008)
6. Musketeer Consortium: Benefits of federated learning – manufacturing industry specialists give their thoughts (automotive sector). Medium Blog (2022). https://h2020musketeer.medium.com/benefits-of-federated-learning-manufacturing-industry-specialists-give-their-thoughts-automotive-87e074b786ae. Accessed 10 July 2025
7. Islam, F., Raihan, A.S., Ahmed, I.: Applications of federated learning in manufacturing: identifying the challenges and exploring the future directions with Industry 4.0 and 5.0 visions. In: Rahman, M.M. (ed.) Proceedings of 8th North American Conference on Industrial Engineering and Operations Management. IEOM, Houston, pp. 1–9 (2023)
8. Li, Q., Wen, Z., Wu, Z., et al.: A survey on federated learning systems: vision, hype and reality for data privacy and protection. IEEE Trans. Knowl. Data Eng. **35**, 3347–3366 (2023). https://doi.org/10.1109/TKDE.2021.3124599
9. Agrawal, S., Sarkar, S., Aouedi, O., et al.: Federated learning for intrusion detection systems: concepts, challenges and future directions. Comput. Commun. **195**, 346–361 (2022). https://doi.org/10.1016/j.comcom.2022.09.012
10. Yang, Q., Liu, Y., Chen, T., et al.: Federated machine learning: concept and applications. ACM Trans. Intell. Syst. Technol. **10**(2), 12 (2019)
11. Zhang, J., Guo, S., Qu, Z., et al.: Adaptive vertical federated learning on unbalanced features. IEEE Trans. Parallel Distrib. Syst. **34**, 2074–2085 (2023). https://doi.org/10.1109/TPDS.2022.3178443
12. Verma, D.C.: Federated AI for Real-World Business Scenarios. CRC Press, Boca Raton (2021)
13. Reinhardt, E.M.: How Siemens mastered federated learning with Katulu. Katulu White Paper, Hamburg (2023). https://www.katulu.io/
14. Mehta, M., Bimrose, M.V., McGregor, D.J., et al.: Federated learning enables privacy-preserving and data-efficient dimension prediction and part qualification across additive manufacturing factories. J. Manuf. Syst. **74**, 752–761 (2024). https://doi.org/10.1016/j.jmsy.2024.04.031
15. Qu, X., Wang, S., Hu, Q., Cheng, X.: Proof of federated learning: a novel energy-recycling consensus algorithm (2021). arXiv:2101.04026
16. Liu, H., Li, S., Li, W., et al.: Efficient decentralized optimization for edge-enabled smart manufacturing: a federated learning-based framework. Future Gener. Comput. Syst. **157**, 422–435 (2024). https://doi.org/10.1016/j.future.2024.03.043
17. Panahifar, F., Byrne, P.J., Salam, M.A., et al.: Supply chain collaboration and firm performance. J. Enterp. Inf. Manag. **31**, 358–379 (2018). https://doi.org/10.1108/JEIM-08-2017-0114
18. Zheng, G., Ivanov, D., Brintrup, A.: An adaptive federated learning system for information sharing in supply chains. Int. J. Prod. Res. **63**, 1–23 (2025). https://doi.org/10.1080/00207543.2024.2432469
19. Wang, Y., Peng, H., Su, Z., et al.: A platform-free proof of federated learning consensus mechanism for sustainable blockchains. SSRN 4157680 (2022)

20. Yang, Z., Chen, M., Saad, W., et al.: Energy-efficient federated learning over wireless communication networks. IEEE Trans. Wirel. Commun. **20**, 1935–1949 (2021). https://doi.org/10.1109/TWC.2020.3037554
21. Qu, X., Wang, S., Hu, Q., et al.: Proof of federated learning: a novel energy-recycling consensus algorithm. IEEE Trans. Parallel Distrib. Syst. **32**, 2074–2085 (2021). https://doi.org/10.1109/TPDS.2021.3056773
22. Haddod, F., Dingli, A.: Intelligent digital twin system in the semiconductors manufacturing industry. In: Artificial Intelligence in Industry 4.0, vol. 928. Lecture Notes in Electrical Engineering, pp. 99–113 (2021). https://doi.org/10.1007/978-3-030-61045-6_8
23. Dritsas, E., Trigka, M.: Federated learning for IoT: a survey of techniques, challenges, and applications. J. Sens. Actuator Netw. **14**, 9 (2025). https://doi.org/10.3390/jsan14010009
24. Lakshminarayan, K., Harp, S.A., Samad, T.: Imputation of missing data in industrial databases. Appl. Intell. **11**, 259–275 (1999). https://doi.org/10.1023/A:1008334909089
25. Zhang, D.: Inconsistencies in big data. In: Proc. 12th IEEE International Conference on Cognitive Informatics & Cognitive Computing, pp. 61–67. IEEE, New York (2013)
26. Gomes, V.B.F., Kleppmann, M., Mulligan, D.P., et al.: Verifying strong eventual consistency in distributed systems. Proc. ACM Program. Lang. **1**(OOPSLA), 1–28 (2017). https://doi.org/10.1145/3133933
27. Zhang, L., Zhu, T., Xiong, P., et al.: A game-theoretic federated learning framework for data quality improvement. IEEE Trans. Knowl. Data Eng. **35**, 10952–10966 (2023). https://doi.org/10.1109/TKDE.2022.3230959
28. Xu, P., Ji, X., Li, M., et al.: Small-data machine learning in materials science. NPJ Comput. Mater. **9**, 30 (2023). https://doi.org/10.1038/s41524-023-01000-z
29. Brecko, A., Kajati, E., Koziorek, J., et al.: Federated learning for edge computing: a survey. Appl. Sci. **12**, 9124 (2022). https://doi.org/10.3390/app12189124
30. Roschewitz, D., Hartley, M.-A., Corinzia, L., et al.: IFedAvg: interpretable data-interoperability for federated learning (2021). arXiv:2107.06897
31. Uchihira, N., Ishimatsu, H., Inoue, K.: IoT service business ecosystem design in a global competitive and collaborative environment. In: Proceedings of PICMET 2016. IEEE, Honolulu, pp 1195–1201 (2016)
32. Radanliev, P., De Roure, D.C., Nurse, J.R.C., et al.: Supply chain design for the industrial internet of things and Industry 4.0. University of Oxford (2019)
33. Schuh, G., Potente, T., Thomas, C., et al.: Approach for reducing data inconsistencies in production control. In: Zäh, M. (ed.) Enabling Manufacturing Competitiveness and Economic Sustainability (CARV 2013), pp. 347–351. Springer, Cham (2014)
34. Fernandez, J.D., Brennecke, M., Barbereau, T., et al.: Federated learning: organizational opportunities, challenges and adoption strategies. IEEE Softw. **40**, 87–95 (2023)
35. Rong, K., Hu, G., Lin, Y., et al.: Understanding business ecosystem using a 6C framework in IoT-based sectors. Int. J. Prod. Econ. **159**, 41–55 (2015). https://doi.org/10.1016/j.ijpe.2014.09.003
36. Otto, B., Jarke, M.: Designing a multi-sided data platform: findings from the international data spaces case. Electron. Mark. **29**, 561–580 (2019). https://doi.org/10.1007/s12525-019-00362-x
37. Bonura, S., De Carbonare, D., Díaz-Morales, R., et al.: Increasing trust for data spaces with federated learning. In: Curry, E., Scerri, S., Tuikka, T. (eds.) Data Spaces: Design, Deployment and Future Directions, pp. 89–106. Springer, Cham (2022)
38. Gou, J., Hu, Y., Sun, L., et al.: Collaborative knowledge distillation via filter knowledge transfer. Expert Syst. Appl. **238**, 121884 (2024). https://doi.org/10.1016/j.eswa.2023.121884
39. Arazzi, M., Cihangiroglu, M., Nicolazzo, S., et al.: Secure federated data distillation (2025). arXiv:2403.01234
40. Xu, Z., Wang, J., Xu, X., et al.: A survey of reinforcement learning-driven knowledge distillation: techniques, challenges, and applications. Preprints.org 202503.0903 (2025). https://doi.org/10.20944/preprints202503.0903.v2
41. Kembuegler, A.: Federated learning for trusted data spaces. International Data Spaces Association. Dortmund (2020)

42. Navia-Vázquez, A., Vázquez, M.A., Cid-Sueiro, J.: "A priori" Shapley data value estimation for risk-balanced data monetization in federated learning. Neurocomputing **463**, 129–143 (2021)
43. Xu, Y., Chen, M.: An internet of things-based framework to enhance just-in-time manufacturing. Proc. Inst. Mech. Eng. B **232**, 2353–2363 (2018). https://doi.org/10.1177/0954405417731467
44. Kallisch, J., Wunck, C.: Development of a prototype for a process support and analysis platform for SMEs. EPiC Ser. Comput. **31**, 20–33 (2020)
45. GAIA-X Foundation: Supply chain collaboration in a connected industry. GAIA-X use-case report, Brussels (2021). https://www.data-infrastructure.eu/GAIAX/Redaktion/EN/Artikel/UseCases/supply-chain-collaboration-in-a-connected-industry.html. Accessed 10 July 2025
46. Kallisch, J., Niemann, K.-H., Wunck, C., et al.: Securing data sovereignty and data security for independent participants in supply chains. OFFIS Research Report 01–2025, Oldenburg (2025)
47. Kallisch, J., Denkena, B., Kramer, K., et al.: Zukunftslabor Produktion. ZWF **119**, 372–377 (2024). https://doi.org/10.1515/zwf-2024-1061
48. Chen, S., Shi, R., Ren, Z., et al.: A blockchain-based supply chain quality management framework. In: Proc. 14th IEEE International Conference on e-Business Engineering (ICEBE). IEEE, Shanghai, pp. 172–176 (2017)
49. Zhang, C., Zhou, G., He, J., et al.: A data- and knowledge-driven framework for digital twin manufacturing cell. Proc. CIRP **83**, 345–350 (2019). https://doi.org/10.1016/j.procir.2019.04.084
50. Bonawitz, K., Ivanov, V., Kreuter, B., et al.: Practical secure aggregation for privacy-preserving machine learning. In: Proceedings of 2017 ACM SIGSAC Conference on Computer and Communications Security. ACM, New York, pp. 1175–1191 (2017)
51. Yin, X., Zhu, Y., Hu, J.: A comprehensive survey of privacy-preserving federated learning. ACM Comput. Surv. **54**, 1–36 (2022). https://doi.org/10.1145/3460427
52. Konečný, J., McMahan, H.B., Yu, F.X., et al.: Federated learning: strategies for improving communication efficiency (2017). arXiv:1610.05492
53. Paragliola, G., Coronato, A.: Definition of a novel federated learning approach to reduce communication costs. Expert Syst. Appl. **189**, 116109 (2022). https://doi.org/10.1016/j.eswa.2021.116109

Chapter 4
Large Multimodal Model-Based Simulation Game as a Socio-Technical System for Value Stream Analysis and Design

Mick Geisthardt, Lutz Engel, and Jorge Marx Gómez

Abstract In the production of capital and consumer goods, human improvement teams are challenged with optimizing value creation while ensuring resource efficiency. The established method of value stream analysis and design, rooted in lean management, has evolved to include material flow cost accounting, information logistics, and external influencing factors. These expansions significantly increased data complexity, rendering traditional pen-and-paper approaches impractical. While existing research emphasizes analyzing current value streams using real-time data and ad-hoc optimization, structured methods for value stream design and implementation strategies remain underdeveloped. To address this gap, this chapter utilizes a novel engineering and management approach—the CRISP-DM-based gamification framework. Building on this framework, a socio-technical simulation game named "North Star—The Value Stream Game" was developed, validated, and deployed, emphasizing interaction between humans and digital systems. The simulation game supports human improvement teams across both initial education and professional development, facilitating knowledge acquisition, practical skills development, and enhanced self-assessed competence. Participants can progress through defined levels of mastery from foundational understanding (level 1) to training proficiency (level 5). Furthermore, integrating large multimodal models (LMMs) enables interactive collaboration between humans and digital systems within the simulation game. Empirical results confirm the educational effectiveness of combining gamification, CRISP-DM, and LMMs, significantly reducing complexity barriers in value stream analysis and design after just one intervention. Statistically significant improvements were observed in knowledge acquisition ($p = 0.002$), practical skills

M. Geisthardt (✉) · J. Marx Gómez
Department of Computing Science, Carl von Ossietzky University of Oldenburg, Oldenburg, Germany
e-mail: mick.geisthardt@jade-hs.de

M. Geisthardt · L. Engel
Department Management, Information, Technology, Jade University of Applied Sciences, Wilhelmshaven, Germany

M. Mora et al. (eds.), *Engineering and Management of Data Science, Analytics, and AI/ML Projects*, Intelligent Systems Reference Library 282,
https://doi.org/10.1007/978-3-032-06889-7_4

development ($p < 0.001$), and self-assessed competence ($p = 0.038$). Findings highlight the potential of socio-technical simulation games for teaching complex methods like value stream analysis and design in a resource-constrained world.

Keywords CRISP-DM-based gamification framework · Socio-technical simulation game · Initial education and professional development · Large Multimodal Models (LMMs) · Value stream analysis and design

4.1 Introduction

In the production of industrial goods—ranging from capital-intensive machinery to consumer-oriented products like food—improvement teams consistently face the challenge of maximizing value creation while ensuring resource efficiency. For decades, lean management tools such as value stream analysis and design have been widely utilized to identify and eliminate inefficiencies within internal production and logistic processes [1, 2]. Originally, these methods focused on traditional metrics such as lead times, cycle times, and inventory reduction. However, recent methodological advancements have expanded this scope, incorporating new aspects like material flow cost accounting [3], information logistics [4–6], and external influencing factors such as market dynamics, digitalization, and sustainability policies [7]. These expansions have unlocked new opportunities but also have introduced new complexity, transforming value stream analysis and design into a highly complex and data-intensive task.

As the methodology diversifies, traditional pen-and-paper or spreadsheet-based approaches quickly became inadequate. Consequently, human improvement teams require new solutions that not only assist in managing this new complexity but also fundamentally transform how teams develop foundational understanding, practical application, mastery, and even training proficiency in complex methods like value stream design [8–11].

Existing research and practical solutions primarily focus on approaches, placing significant emphasis on the analysis and simulation of current value streams. Such methods frequently integrate real-time data analytics and ad-hoc optimization features [12–21]. Despite these advancements, a crucial area remains underdeveloped: the design of target value streams in the direction of the ideal state [22]. Specifically, there is very limited research into methods or tools for value stream design, i.e., identifying resource-efficient target value streams and developing corresponding implementation strategies for optimization.

To bridge this research gap, this chapter utilizes a novel engineering and management approach that integrates gamification with the Cross-Industry Standard Process for Data Mining (CRISP-DM) —called the CRISP-DM-based gamification framework. The core objective of this approach is to enable initial education and professional development of human improvement teams teaching to identify and design resource-efficient target value streams, effectively navigating real-world complexity.

Grounded in this integration, a socio-technical simulation game titled "North Star—The Value Stream Game" [23–25] was developed, validated, and deployed. The game specifically addresses two critical dimensions of resource efficiency: first, it educates human improvement teams to design value streams that inherently optimizes resource consumption; second, it educates to adopt resource-efficient thinking as a central component of their strategic mindset.

Furthermore, this chapter explores the integration and application of large multimodal models (LMMs)—a class of artificial intelligence (AI) that combines multiple data modalities, such as text, images, numerical data, and domain-specific knowledge [26]. By leveraging LMMs within the socio-technical simulation game, human improvement teams engage collaboratively with a digital system [27], facilitating interactive learning, advanced analytical support, and intelligent guidance throughout the complex process of value stream design.

Experimental empirical results confirm the educational effectiveness of combining gamification, CRISP-DM, and LMMs while significantly lowering barriers typically associated with mastering complex analytical tasks like value stream design after just one intervention. Quantitative assessments using inferential statistics methods [28, 29] indicate statistically significant improvements across the three dimensions: knowledge acquisition ($p = 0.002$), practical skills development ($p < 0.001$), and self-assessed competence ($p = 0.038$).

These findings highlight the strong potential of the approach to significantly enhance the effectiveness of both initial education and ongoing professional development. Drawing on the five levels of mastery from lean management—ranging from Level 1 (awareness) and Level 2 (basic understanding), through Level 3 (independent application) and Level 4 (consistent, effective use), up to Level 5 (ability to teach others) [11]—the use of socio-technical simulation games proves to be a promising method for teaching complex concepts such as value stream design, particularly in a resource-constrained world.

Following this introduction, the chapter first presents a theoretical overview of value stream analysis and design highlighting the evolving complexities that challenges human improvement teams. Subsequently, the conceptual and methodological foundations of socio-technical systems, game-based learning with AI, and LMMs are briefly explained (Sect. 4.2). Thereafter, the chapter details the applied methodology (Sect. 4.3), followed by implementation, analysis, and discussion of empirical results, focusing on the practical implications of the presented socio-technical simulation game and its evaluation as tool for initial education and professional development (Sect. 4.4 and 4.5). Finally, the chapter concludes by summarizing key insights and offering an outlook on directions for future research (Sect. 4.6).

4.2 Background

This Sect. 4.2 presents the theoretical background necessary for comprehending the chapter's core contributions. It briefly introduces key concepts, including value stream analysis and design, socio-technical systems, game-based learning with AI, and LMMs, laying the basis for the results, discussions, and conclusions detailed in Sects. 4.3, 4.4, and 4.5.

4.2.1 Value Stream Analysis and Design

An end-to-end value stream can be defined as the complete sequence of value-adding and non-value-adding activities that are essential to deliver a product or product family from initial sourcing through manufacturing processes to the final customer [1]. For example, in the production of photovoltaic modules, the value stream starts from sourcing raw silicon material, proceeds through wafer production (P1), continues with cell manufacturing (P2), and concludes with module assembly (P3), eventually reaching the end customer. As illustrated in Fig. 4.1, the abstract concept of a value stream—or its visual representation—can be broadly divided into three sections. The first, upper section represents suppliers, customers, and information related to the flow of data from the areas of production planning, control, and communication. The second, middle section includes the processes themselves, material flow, and inventories. The third, lower section serves to evaluate the value stream from a temporal perspective, aiming to determine both the lead time and the value-adding processing time.

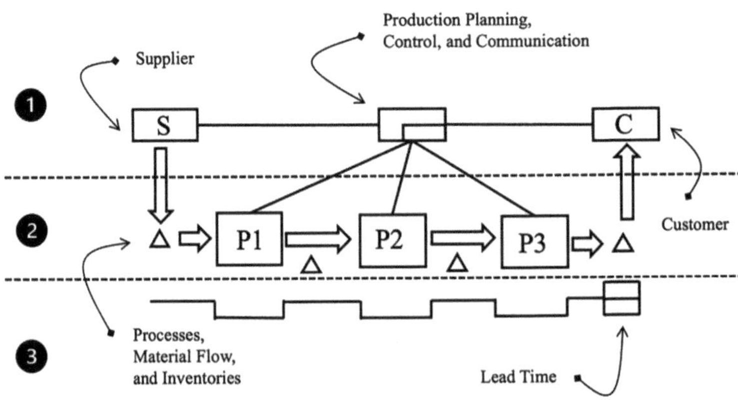

Fig. 4.1 Schematic representation of a value stream divided into three sections: (1) information flow including suppliers, customers, and production planning and control; (2) operational flow including processes, material flow, and inventories; and (3) time analysis including lead time

Throughout these sequential processes, both value-adding activities (e.g., precise cell coating during cell manufacturing in P2) and non-value-adding activities (e.g., waiting times or excess inventory between wafer production (P1) and cell manufacturing (P2)) occur. Identifying and systematically reducing these non-value-adding activities can significantly enhance overall value creation enhance resource efficiency.

Applying traditional value stream analysis involves first mapping the current situation of photovoltaic module production using standardized visual elements, typically illustrated on whiteboards or paper [1]. Subsequently, human improvement teams identify inefficiencies using core principles of lean management, e.g., Muda, Mura, and Muri [30]:

- Muda (waste) refers to activities consuming resources without customer-perceived value, such as excess wafer inventories between wafer production (P1) and cell manufacturing (P2).
- Mura (unevenness) represents fluctuations or inconsistencies, for instance, production speed variations between wafer production and cell manufacturing, causing bottlenecks.
- Muri (overburden) indicates excessive strain placed on employees or equipment, such as unrealistic workload expectations during module assembly (P3), potentially resulting in quality issues or fatigue.

Following this analytical step, improvement teams utilize established guidelines for value stream design to create optimized target value streams. These designs aim for enhanced resource efficiency, shorter lead times, and waste minimization. The overarching goal of value stream design is derived from the Toyota Production System (TPS), represented as the concept of an ideal state, often referred to as the "north star" [22]. This ideal state emphasizes zero defects, 100% value creation, one-piece flow aligned with customer demand, and ensuring safety for employees. Lastly, the newly developed value stream design is implemented, validated, and standardized in the real-world value stream.

In recent years, value stream analysis and design has evolved significantly, incorporating broader methodological aspects to address increasing complexity. While traditional value stream analysis primarily focuses on lead-time metrics involving suppliers (S), processes (P), inventories (I), customers (C), production planning and control (PPC), and lead time (LT), advanced methods now integrate additional aspects. Specifically, advanced value stream analysis considers material flow cost accounting [3], information logistics [4], and external influencing factors [7], enabling a more comprehensive and realistic depiction of production complexity.

Figure 4.2 illustrates a current value stream that incorporates these advanced methodological expansions. Traditional lead time-centered elements are depicted in black.

Addressing the limitation of focusing exclusively on lead time, material flow cost accounting (according to DIN EN ISO 14051:2011) [31] systematically evaluates inputs and outputs of processes, distinguishing between product input (PI), value-adding product output (PO), and non-value-adding product output (NPO), depicted

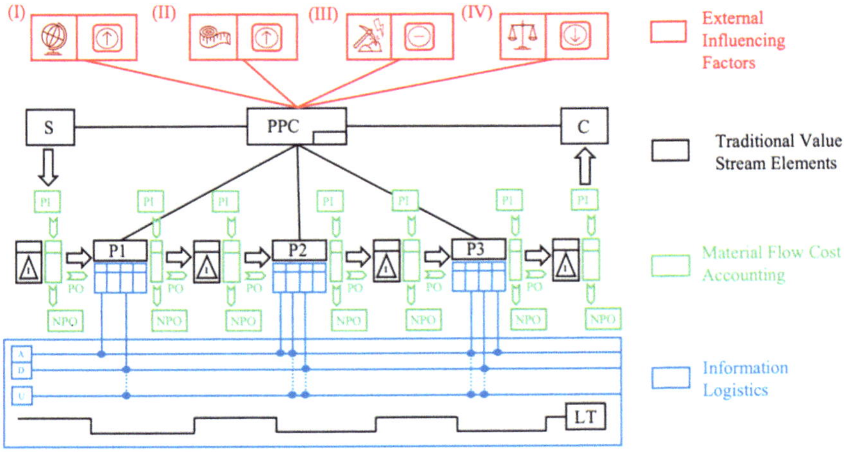

Fig. 4.2 Example of a value stream in the current situation, enhanced through material flow cost accounting, information logistics, and external influencing factors to illustrate the increasing complexity of modern value stream analysis and design practices

in green. For example, in wafer production (P1), silicon ingots constitute the product input (PI). The correctly produced silicon wafers represent the product output (PO), whereas damaged wafers or silicon dust generated during slicing constitute non-value-adding product output (NPO).

These categories facilitate detailed analysis of material, energy, system, and waste management costs, providing key performance indicators such as absolute monetary value creation or the degree of value creation for in-depth analysis [3].

Furthermore, incorporating information logistics expands the analytical scope to include activities related to information flows and their key performance indicators, depicted in blue. Information logistics tracks analog (A) and digital (D) storage media, along with their utilization (U). For example, digital data tracking systems may manage wafer inventory and quality records (P1), while assembly instructions in module assembly (P3) could partially rely on analog documentation printed on paper and distributed through human channels, highlighting opportunities for digitization. Typical evaluation metrics include digitization rate, data availability, and data utilization [6].

Finally, exclusively internal analyses may not fully account for risks from external influencing factors. External influencing factors—represented in red—such as geopolitical conditions of suppliers (I), procurement market risks (II), energy availability (III), and regulatory environments (IV) can severely impact production efficiency. For instance, geopolitical instability affecting silicon supply chains could disrupt wafer production (P1), while regulatory changes in energy policy may influence demand patterns, impacting module assembly (P3). Integrating these external influencing factors into value stream analyses enables a more holistic view, promoting comprehensive risk management and optimization [7].

While the methodological advancements outlined above significantly extend the complexity and analytical depth of value stream analysis and design, the developments, analyses, and evaluations conducted within this chapter initially focus on teaching and learning the principles of traditional value stream design. Specifically, emphasis is placed on systematically designing target value streams and enhancing resource efficiency. When applying the simulation game within a business context, the value streams observed in the game can be improved, and measures derived from the developed value stream design can be implemented in the real world. The presented concepts and methods are suitable for both initial education and professional development, enabling learners to systematically progress through clearly defined levels of mastery, ranging from foundational understanding and practical application to mastery and even training proficiency in value stream design.

4.2.2 Socio-Technical Systems

Understanding value stream design, along with its increasing methodological complexity, naturally leads to questions about how human improvement teams can practically apply this method in realistic settings. Addressing this requires consideration of both humans and technology and their interaction within organizational contexts. Thus, the concept of socio-technical systems becomes particularly relevant, offering insights into effectively managing this interplay in initial education and professional development of resource-efficient value stream design.

As illustrated in Fig. 4.3, the simulation game "North Star—The Value Stream Game" specifically addresses these considerations by systematically supporting and guiding the initial education and professional development of human improvement teams in designing target value streams (task).

Human improvement teams become trainees and players (human) and interact with an LMM (technology)—specifically "NorthStarGPT"—integrated as a dynamic, context-sensitive support and guidance system. Within a clearly defined, iterative simulation game (structure), teams analyze predefined value stream scenarios and resource constraints (input) to generate optimized actionable improvement measures (value stream design) and enhanced competence in resource-efficient decision-making (output).

In industrial practice, it has long been recognized that successfully improving value streams requires more than pure technology. The concept of socio-technical systems captures this insight by emphasizing the integrated consideration of both technical and human components within organizational systems [27]. Socio-technical systems emerged as a reaction to the realization that optimizing technology alone rarely leads to sustainable improvements unless human and social factors are equally considered [32].

Thus, the core idea behind socio-technical systems is the joint optimization of social and technical elements to achieve lasting effectiveness and resilience

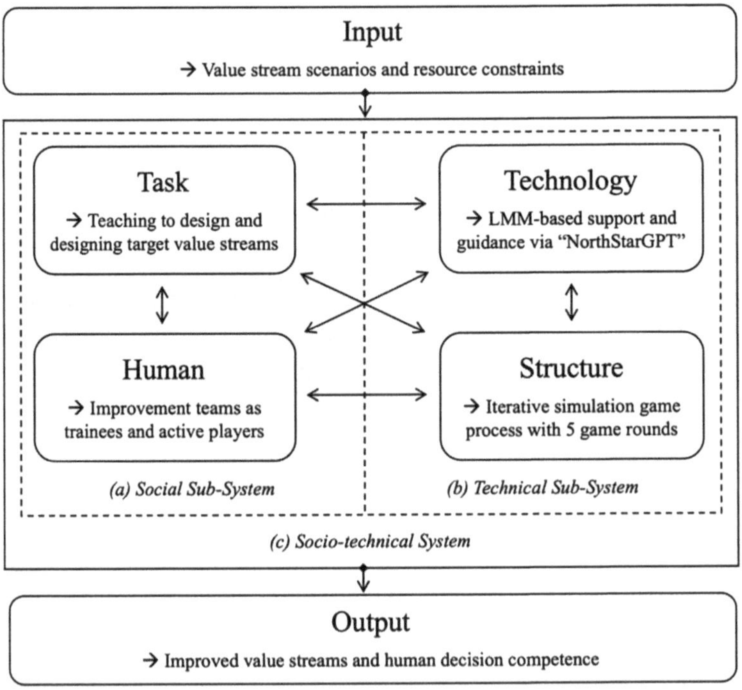

Fig. 4.3 Interaction of technical (**a**) and social (**b**) sub-systems within the socio-technical system (**c**) of the simulation game "North Star—The Value Stream Game"

within organizations. Rather than treating technical efficiency and employee productivity as separate goals, joint optimization seeks synergies between the two [33]. Consequently, socio-technical systems are inherently systemic: they recognize mutual dependencies and interactions between technology, organizational structures, processes, and human behavior [34].

Applying socio-technical systems is particularly valuable within the context of value stream design. As described previously, increasing methodological complexity—such as integrating material flow cost accounting, information logistics, and external influencing factors—places greater demands on the analytical skills and collaborative abilities of human improvement teams. Here, a purely technological or purely organizational approach would fall short. Instead, human improvement teams require a system that enable effective integration of digital tools while actively fostering the human skills necessary to handle complex decision-making tasks.

Implementing socio-technical systems in practice is not straightforward. Typical challenges include a tendency toward technocentric solutions that neglect human factors, insufficient involvement of employees in the design process, or gaps in leadership commitment to holistic system improvements [35].

Despite these potential challenges, socio-technical systems offer significant advantages—particularly for human improvement teams involved in learning-intensive activities such as value stream design. Meaningful socio-technical integration of digital and smart working solutions significantly enhances sustainability, adaptability, and overall effectiveness in complex organizational environments [36].

The implementation of methods, such as simulation games enhanced by AI, as a socio-technical system enables humans to effectively learn resource efficient decision-making within realistic but controlled scenarios. This is precisely the context in which the methodological framework and simulation game presented in this chapter, "North Star—The Value Stream Game", operates. Here, socio-technical systems serve as a foundation for developing an innovative educational tool, combining human-centered gamification with LMMs. In this context, the term socio-technical simulation game is introduced and defined for the first time:

A **socio-technical simulation game** is a structured, interactive learning format that integrates social elements—such as human decision-making, collaborative problem-solving, competition, communication, and organizational dynamics—with technical components, including simulation models, data-driven systems, and AI-based tools. Specifically designed to facilitate both initial education and professional development, the game positions trainees as active players who progressively build competencies through systematic knowledge acquisition, focused practical skills development, and enhanced self-assessed competence. Within a controlled yet immersive environment, learners collaboratively engage with realistic, complex scenarios, enabling them to effectively understand, apply, and master acquired knowledge and skills.

Unlike using purely digital systems, this concept explicitly incorporates the interplay between humans and technology by embedding realistic and complex problem scenarios into a structured, gamified setting. Trainees are not only technically supported but are actively encouraged to solve challenges collaboratively, which strengthens both social-organizational and technological competence capabilities.

While the concept builds on established theories of socio-technical systems and simulation games, the combination of these elements into a unified, interactive socio-technical simulation game is rarely discussed in current research. Existing literature on socio-technical systems—such as Baxter and Sommerville [33], Bednar and Welch [36], and Sarker et al. [35]—highlights the importance of integrating human and technological perspectives but does not explicitly define or explore this convergence as a distinct learning concept. Therefore, this chapter proposes the socio-technical simulation game as a new conceptual and methodological category, specifically designed to address the demands of complex learning contexts such as value stream design.

4.2.3 Game-Based Learning with Artificial Intelligence

Building on the understanding of socio-technical systems and socio-technical simulation games addressing the interplay between humans and technology, it becomes clear that learning formats aiming to support and guide human improvement teams should reflect this duality. One promising approach that operationalizes socio-technical systems in a practical and engaging way is game-based learning with AI.

Game-based learning with AI combines interactive, motivational learning environments with the adaptive and analytical capabilities of AI. This approach aims to enhance engagement, support critical thinking and problem-solving, and improve long-term knowledge retention through personalized and dynamic learning experiences. By integrating core game mechanics—such as challenge, feedback, and progression—with AI, humans are guided through interactive scenarios that adapt to their individual performance and learning needs [37].

Key components of game-based learning with AI include personalization, adaptive feedback, dynamic content generation, and data-driven assessment [38]. AI algorithms continuously monitor human behavior during gameplay and respond with timely cues, explanations, or guidance when learners face difficulties or make errors. This form of adaptive feedback fosters a growth-oriented learning environment and encourages continuous improvement through targeted support [39].

In addition, AI as digital system can autonomously generate diverse and contextually relevant content—such as new levels, tasks, or decision-making challenges—tailored to the human's current skill level and progress. This adaptability helps maintain humans attention and motivation over extended periods, reducing repetition and cognitive fatigue [40].

Beyond real-time interaction, AI-driven analytics enable the systematic evaluation of human performance. These insights support educators by identifying strengths, knowledge gaps, and learning patterns, thereby facilitating the continuous refinement of content and instructional design [41].

Overall, the integration of AI and game-based learning creates an interactive, responsive, and human-centered educational environment. In the context of this chapter, these principles are applied within the socio-technical simulation game "North Star—The Value Stream Game", enabling human improvement teams to engage with complex value stream design challenges in a guided, dynamic, and personalized way.

4.2.4 Large Multimodal Models

Building on the potential of game-based learning with AI and socio-technical simulation games, it is essential to examine the underlying technologies that enable intelligent, adaptive, and context-aware interactions. A key component in this regard are large language models (LLMs), which form the foundation for many recent advances in natural language processing (NLP) and AI-supported and guided learning environments.

LLMs are a class of machine learning models specifically designed to process, interpret, and generate human language. Trained on vast amounts of textual data, these models learn to recognize complex linguistic patterns and semantic relationships. A central architectural element of modern LLMs is the transformer, which allows for the efficient processing of contextual information across long text sequences. This architecture plays a crucial role in various NLP tasks such as machine translation, text summarization, and question answering. Transformer-based models, including OpenAI's GPT-3 with its 175 billion parameters, have significantly shaped the current landscape of NLP research and applications [42, 43].

In recent years, LLMs have found widespread adoption across various industrial domains. Organizations increasingly leverage these models to automate and optimize text-centric tasks. Common use cases include AI-driven chatbots and virtual assistants in customer support, which can interpret user inputs and provide contextually appropriate responses in multiple languages [44]. In the field of data analysis, LLMs enable the extraction and structuring of relevant information from unstructured text sources—a task that is becoming increasingly important due to the exponential growth of digital data in organizational environments [45]. The industrial relevance of LLMs continues to grow in parallel with advances in computing infrastructure and the availability of big data, which allow for the development and deployment of increasingly powerful and specialized models [46].

A more recent development is the emergence of large multimodal models (LMMs), which expand the capabilities of traditional LLMs beyond text. Multimodality refers to the ability of models to simultaneously process and integrate different types of data—such as text, images, audio, or video. While early LLMs focused solely on linguistic data, recent research efforts have led to the creation of models capable of combining and contextualizing information from multiple sources. This integration enables a deeper and more holistic understanding of complex scenarios, particularly in environments where language interacts with visual, auditory, or contextual cues [47].

In the context of this chapter, LMMs provide the technical foundation for the socio-technical simulation game. Their capacity to understand multimodal content enables more realistic, flexible, and adaptive learning scenarios—particularly when embedded in game-based learning settings aimed at improving resource-efficient value stream design.

4.3 Methodology

To address the increasing complexity of value stream design and to support and guide improvement teams in acquiring the necessary analytical and conceptual competencies, this chapter presents a structured approach for the development and validation of a socio-technical simulation game. The objective of this game, "North Star—The Value Stream Game", is to perform initial education and professional development enabling human improvement teams to develop foundational understanding, practical application, mastery, and even training proficiency of resource efficient value stream design. The simulation game aims to replace passive instructional formats with an active, exploratory, and practice-oriented learning experience. Participants engage in iterative design decisions under realistic constraints, while collaborating with an embedded LMM-based digital system that provides adaptive feedback and analytical support throughout the simulation game.

To ensure that the simulation game is both educational effective and practically applicable, the development process was guided by principles from agile development, the lean start-up methodology, and the concept of the minimum viable product [48, 49].

These principles emphasize iterative refinement, user-centered design, and early validation through real-world feedback. Against this background, a tailored CRISP-DM-based gamification framework was developed and applied. This framework merges the established six-step CRISP-DM [50]—a widely used standard process for data science projects—with the six steps of gamification as proposed by Werbach and Hunter [51]. The result is a cyclical, IT-oriented model that enables a consistent and flexible development process for simulation games, from conceptualization to deployment (see Fig. 4.4).

The application of the CRISP-DM-based gamification framework begins with the phase of business understanding (1), in which the real-world challenges and learning objectives of human improvement teams are analyzed. In this phase, the goal is to define the simulation game's purpose within the broader context of value stream design and to identify key aspects of resource efficiency, data complexity, and team-based problem solving as critical learning dimensions. In the second phase, game understanding (2), these objectives are translated into game-specific learning outcomes. This includes identifying desired user behaviors within the simulation game, defining the player personas, and mapping the learning objectives to game mechanics.

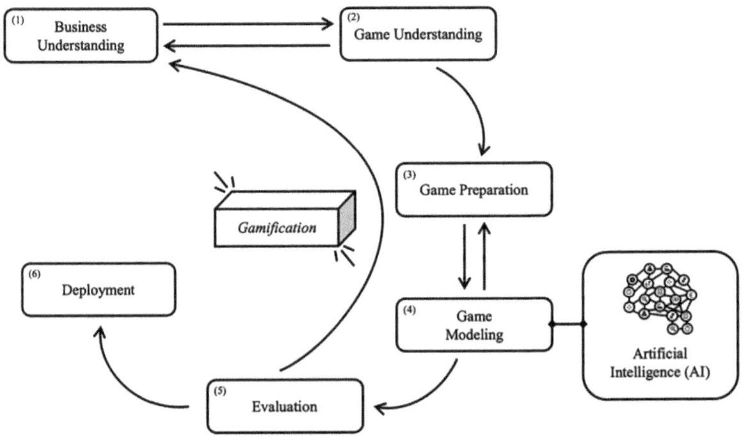

Fig. 4.4 CRISP-DM-based gamification framework combining CRISP-DM and gamification principles to guide the iterative development of a socio-technical simulation game for value stream analysis and design

In the third phase, game preparation (3), the structural and narrative framework of the simulation game is conceptualized. The preparation includes the design of activity cycles, progression stairs, and decision pathways that reflect the complexity of real-world value stream design. This phase also involves defining the integration points for AI components and multimodal user interaction. In the subsequent game modeling phase (4), creative and technical modeling tools are used to translate the conceptual framework into a functional prototype. Key focus areas include AI response logic, interaction, and the visualization key performance indicators within the simulation game.

After prototype development, the evaluation phase (5) involves a series of internal and external validation procedures. If the simulation game meets the defined objectives, it proceeds to the final deployment phase (6).

To evaluate the educational effectiveness of the socio-technical simulation game in detail, a randomized controlled pretest–posttest design was applied. Participants were randomly assigned to either an experimental group (players), which engaged with the simulation game "North Star—The Value Stream Game", or to a control group (observers), which completed the same test instruments without any intervention. Both groups participated in identical pre- and post-assessments, allowing for a structured comparison of learning outcomes. The assessment instruments included:

- Five knowledge-based questions to evaluate theoretical understanding,
- One practical task to assess applied skills in value stream design,
- Three self-assessment questions based on the five levels of mastery framework [11] to measure individual competence development, and
- Additional questions on the gameplay experience, designed to capture subjective engagement and perceived learning benefit.

The study was conducted as part of the university course "Digital Lean Manufacturing" at the Laboratory for Production Management and Technology at Jade University of Applied Sciences, involving 11 participants. Among the participants were 3 female and 8 male students, all of whom were bachelor's students in their fifth semester. Most of them had already gained practical experience in the industry. The topic of value stream analysis and design was already familiar to the participants prior to the intervention.

Participants were randomly assigned to one of two groups: an experimental group (players), which engaged with the simulation game "North Star—The Value Stream Game", or a control group (observers), which did not participate in the gameplay. The implemented randomized controlled pretest–posttest field experiment was integrated into the course but participation in the experiment itself was voluntary.

To analyze the collected data, both descriptive statistics and methods of inferential statistics [29] were applied. Specifically, paired sample t-tests [28] were used to evaluate significant changes in test results before and after the intervention within and between groups.

4.4 Implementation

The simulation game "North Star—The Value Stream Game" by Geisthardt et al. [23–25] is specifically developed to transform the complex methodology of value stream analysis and design into an interactive, immersive learning format that promotes systematic knowledge acquisition, practical skills development, and enhanced self-assessed competence in both initial education and professional development contexts.

The simulation game is designed as a socio-technical simulation game (see Sect. 2.3) with a clear educational objective: to enable participants to systematically progress through defined levels of mastery—from foundational understanding and practical application to mastery and even training proficiency—in key principles of value stream design, and to internalize resource-efficient decision-making under realistic constraints. In doing so, the game bridges the gap between theoretical learning and applied practice in industrial process optimization (see Fig. 4.5).

At its core, the simulation game challenges players to holistically optimize a production system by identifying inefficiencies, selecting appropriate improvement strategies, and implementing value stream design guidelines. A central focus lies on resource efficiency, which is operationalized through the strategic management of resource units (RUs). These represent a limited budget of cognitive, material, or organizational resources that must be allocated to improvement actions. Every decision within the game—such as the selection and sequencing of improvement measures—requires players to balance short-term benefits with long-term value creation.

The game proceeds over five structured rounds, in which players analyze a current situation value stream and iteratively improve and design it, receiving feedback from an LMM called "NorthStarGPT". Each round follows a cyclical structure: players select action cards from a randomized physical set of five cards, apply one of these to

(a) (b)

Fig. 4.5 "North Star—The Value Stream Game" as implemented in the course Digital Lean Manufacturing at Jade University of Applied Sciences: **a** overview of game components including game box, action cards, resource units, progression stairs, and evaluation cards; **b** in-game setup with game board and integrated LMM displayed on screen for player interaction

different stages of the value stream, and receive feedback via "NorthStarGPT" based on their impact and resource usage. This structure mirrors the logic of continuous improvement cycles in lean management, such as PDCA (Plan-Do-Check-Act) [52]. Success is not determined solely by the number of actions taken but by their strategic effectiveness and resource efficiency. Players are rewarded for impactful interventions that reduce waste, improve flow, and bring the system closer to the ideal "north star" state—defined by the principles of zero defects, one-piece flow, 100% value creation, and operational stability [22].

A distinguishing feature of the game is the integration of the LMM called "NorthStarGPT" (based on GPT 4o from OpeanAI), which acts as the interactive digital system within the socio-technical simulation game. Depending on the configuration, the LMM fulfills several roles within the game: it digitally initializes game parameters, explains rules, guides players through decision points, systematically evaluates player moves based on its knowledge base, provides constructive feedback, and manages the implementation within the simulated scenario across the five game rounds. The resulting dynamic interaction creates a socio-technical system, where physical human decision-making is continuously supported—and occasionally challenged—by LMM-driven output. The LMM contributes not only to the simulation game's realism but also to its educational effectiveness by offering real-time guidance and clarification.

From a didactic perspective, the simulation game is deliberately designed to create productive friction: players are encouraged to think critically about trade-offs, analyze cause-effect relationships, and iteratively revise their strategies based on performance feedback. The simulation game narrative, resource mechanics, and AI interaction are tightly interwoven to reinforce key learning outcomes. These include systems thinking, lean principles, and the development of decision-making competence under constraints—skills that are central to modern value stream design in industrial practice.

4.5 Results and Discussion

This Sect. 4.5 presents the results of the empirical evaluation and discusses the educational effectiveness of the socio-technical simulation game. It focusses on the analysis of the three core aspects of the evaluation: knowledge acquisition, practical skill development, and self-assessed competence. The evaluation is complemented by user feedback on the gameplay experience and concludes with a discussion of the key findings and their implications.

Table 4.1 Results of the pre- and post-tests on knowledge acquisition for the experimental and control groups ($N = 11$). Values are presented as means with standard deviations (SD). Statistical significance was tested using paired-sample *t*-tests using IBM SPSS Statistics

Group	Pre-test mean (SD)	Post-test mean (SD)	Mean Δ (%)	*p*-value	Cohen's *d*	Significance
Experimental ($n = 6$)	1.3 (0.5)	3.0 (0.9)	+125.0	0.002	0.82	Yes
Control ($n = 5$)	3.0 (1.4)	3.0 (1.2)	±0	0.500	0.70	No

4.5.1 *Knowledge Acquisition*

To assess the effect of the socio-technical simulation game on participants' theoretical understanding of value stream analysis and design, five knowledge-based questions were included in both the pre- and post-tests. Each question presented five possible answers, of which one or more could be correct. The content focused on core concepts such as the objectives of value stream analysis and design, relevant and non-relevant information elements, correct application of pull principles, and the identification of valid and invalid value stream design guidelines. One point was awarded per correctly answered question, resulting in a maximum score of five points.

As summarized in Table 4.1, in the experimental group, which interacted with the simulation game, participants showed a clear improvement in their knowledge scores. The average score before the intervention was 1.3 (standard deviation (SD) = 0.5), increasing to 3.0 (SD = 0.9) in the post-test. This represents a mean improvement of 125.0%, indicating a substantial learning gain. The paired-sample *t*-test, conducted using IBM SPSS Statistics, confirmed the statistical significance of this improvement ($p = 0.002$, cohen's $d = 0.82$), suggesting a large effect size.

In contrast, the control group, which did not participate in the simulation game completed the same assessments, showed no meaningful change. The average pre-test score was 3.0 (SD = 1.4) and remained at 3.0 (SD = 1.2) in the post-test, with no average change. The corresponding *t*-test did not indicate a significant difference ($p = 0.500$, cohen's $d = 0.70$).

The results indicate that the simulation game had a significant and positive effect on participants' theoretical knowledge in the field of value stream analysis and design. The increase in scores within the experimental group, combined with the stability of results in the control group, provides evidence for the educational effectiveness of the socio-technical learning approach.

4.5.2 *Practical Skills Development*

To evaluate the development of practical skills in value stream analysis and design, participants were given an applied task in both the pre- and post-tests. The task required participants to translate a textual description of a simplified production process into a value stream mapping using correct lean management symbols. The process included the stages: supplier → incoming goods → processing → assembly → shipping → customer. Participants had five minutes to design the current situation using standard value stream symbols and to identify potential weaknesses through the placement of kaizen flashes [1].

The task was designed to assess the participants' ability to apply value stream analysis and design conventions in a structured and resource-conscious way. Specifically, the evaluation focused on the correct use of standardized symbols (the "language of lean managers"), the structural logic of the mapping design, and the ability to identify and mark improvement areas. A maximum of 15 points could be achieved. Evaluation was carried out independently by experts from the Institute for Production and Service Systems at Jade University of Applied Sciences, based on a standardized scoring rubric.

As summarized in Table 4.2, in the experimental group, the average score improved from 7.2 (SD = 2.8) to 11.2 (SD = 2.9) after the simulation game, corresponding to a mean increase of 55.8%. The improvement was statistically significant according to the paired-sample t-test ($p < 0.001$, cohen's $d = 1.27$), indicating a strong effect size. The results suggest that the simulation game effectively supported the transfer of theoretical knowledge into applied practice, especially in terms of symbol fluency, structure, and design logic.

Interestingly, the control group, which did not actively participate in the simulation game but observed the gameplay and instructions during the session, also demonstrated a statistically significant improvement. Their average score increased from 5.6 (SD = 3.2) to 10.4 (SD = 3.6), which equals a mean increase of 85.7% ($p < 0.001$, cohen's $d = 1.48$). This finding suggests that even passive exposure to the simulation game—through observation—may foster learning, likely due to the visibility of decision-making processes, terminology, and value stream logic reinforced throughout the gameplay.

It can be observed, that both active and observational engagement with the simulation game led to substantial improvements in the practical application of value stream analysis and design concepts. These results underline the value of socio-technical

Table 4.2 Results of the pre- and post-tests assessing practical value stream analysis and design skills in the experimental and control groups ($N = 11$). Scores are based on a standardized task with a maximum of 15 points. Both groups showed statistically significant improvements; the experimental group actively participated in the simulation game, while the control group observed passively

Group	Pre-test mean (SD)	Post-test mean (SD)	Mean Δ (%)	p-value	Cohen's d	Significance
Experimental ($n = 6$)	7.2 (2.8)	11.2 (2.9)	+55.8	< 0.001	1.27	Yes
Control ($n = 5$)	5.6 (3.2)	10.4 (3.6)	+85.7	< 0.001	1.48	Yes

simulation game, in which structured simulation tasks, visual elements, and guided decision-making processes interact to support applied learning outcomes.

4.5.3 Self-Assessed Competence

To capture participants' subjective perception of their own competence and learning progress, three self-assessment questions were included in both the pre- and post-test. Participants were asked to rate their ability to (1) conduct a value stream analysis, (2) design a target value stream, and (3) apply value stream design guidelines in a resource efficient manner. The self-assessment was based on the five levels of mastery, ranging from level 1 (awareness of the concept), level 2 (basic understanding), level 3 (independent application), level 4 (consistent and effective use), to level 5 (ability to teach others) [11]. This framework encourages learners to reflect on their level of independence, confidence, and sustainability in applying acquired knowledge and skills.

As summarized in Table 4.3, the experimental group participants rated their self-perceived ability to conduct a value stream analysis higher after completing the simulation game. The mean score increased from 2.3 (SD = 0.8) to 3.2 (SD = 1.0), corresponding to a mean increase of 35.7%. Although the *p*-value of 0.071 did not reach the conventional level of statistical significance, the effect size was large (cohen's $d = 1.17$), indicating an improvement in perceived competence.

In the control group, which did not play the game but was present during the session, the mean score concerning value stream analysis increased only slightly from 2.6 (SD = 0.5) to 3.0 (SD = 0.7), representing a 18.8% increase. The *t*-test result ($p = 0.089$, cohen's $d = 0.55$) suggests a moderate effect, but not a statistically

Table 4.3 Pre- and post-test results for self-assessed competence in value stream analysis and design based on the five levels of mastery ($N = 11$). Scores range from 1 (awareness) to 5 (ability to teach others). Improvements were more pronounced in the experimental group, with a statistically significant increase observed for value stream design

Group	Pre-test mean (SD)	Post-test mean (SD)	Mean Δ (%)	*p*-value	Cohen's *d*	Significance
Value stream analysis						
Experimental ($n = 6$)	2.3 (0.8)	3.2 (1.0)	+35.7	0.071	1.17	No
Control ($n = 5$)	2.6 (0.5)	3.0 (0.7)	+18.8	0.089	0.55	No
Value stream design						
Experimental ($n = 6$)	2.7 (0.5)	3.2 (0.7)	+15.4	0.038	0.55	Yes
Control ($n = 5$)	2.8 (0.8)	3.0 (0.7)	+7.1	0.311	0.84	No

significant change. The contrast between groups implies that the simulation game had a positive influence on participants' self-perception regarding their analytical abilities.

Furthermore, as illustrated in Table 4.3, a similar trend was observed in the self-assessment related to value stream design. The experimental group reported an increase from 2.7 (SD = 0.5) to 3.2 (SD = 0.7), representing a 15.4% improvement, with result reaching statistical significance ($p = 0.038$, cohen's $d = 0.55$). Participants demonstrated a greater sense of confidence in designing value streams after the simulation game-based learning experience.

In contrast, the control group showed only a marginal improvement of 7.1%, with the mean increasing from 2.8 (SD = 0.8) to 3.0 (SD = 0.7), and no statistically significant effect was detected ($p = 0.311$, cohen's $d = 0.84$).

The results show that the simulation game contributed not only to objective improvements in knowledge and skills, but also to greater confidence and perceived competence among participants. The experimental group, having actively engaged with the simulation game, consistently rated their level of mastery higher after the intervention. Although the control group also showed minor improvements—likely influenced by passive observation of the simulation—the gains in the experimental group were more substantial and aligned with the learning goals.

4.5.4 User Feedback and Game Experience

To complement the quantitative evaluation, participants were asked to provide feedback on their experience with the simulation game. A total of six items were used to assess perceived learning outcomes, motivation, practical relevance, and game dynamics. Responses were collected on a five-point likert scale, ranging from strong disagreement (1) to strong agreement (5). The feedback was provided by both players and observers.

Initial results show a generally positive reception of the simulation game format. Most participants agreed that the game improved their understanding of value stream analysis and design and supported the practical application of resource efficiency concepts. Players rated the game as motivating and highlighted the collaborative and competitive dynamics as beneficial to their learning experience. Qualitative comments also emphasized the usefulness of real-time feedback and decision consequences.

The overall feedback supports the simulation game's potential to create an engaging and educational effective learning environment and confirms its relevance for competence development in industrial value stream design.

4.6 Summary and Discussion

The results of the empirical evaluation provide a differentiated picture of the learning effects achieved through the socio-technical simulation game "North Star—The Value Stream Game". Across all measured dimensions—knowledge acquisition, practical skills development, and self-assessed competence—the data shows positive developments, particularly within the experimental group that actively engaged with the simulation game. While the control group exhibited minor or moderate learning effects in some areas, likely due to observational exposure, the structured gameplay with interactive decision-making and AI-supported feedback proved more effective overall in promoting measurable and perceived learning outcomes.

In terms of knowledge acquisition, participants in the experimental group showed a statistically significant improvement. The average score increased from 1.3 to 3.0, with a large effect size (cohen's $d = 0.82$). This reflects a clear gain in theoretical understanding of core value stream design principles. In contrast, the control group remained stable, showing no meaningful change.

Regarding practical skills development, both groups showed significant improvement in the post-test performance, but with different implications. The experimental group improved by 55.8%, while the control group—who observed the gameplay without direct interaction—achieved an even higher improvement of 85.7%. These results suggest that the simulation game format supports practical learning not only through active participation but also through observation of decisions, feedback, and systemic effects within the learning environment. It highlights the simulation game's potential to create a shared socio-technical space that benefits both active and passive participants.

In terms of self-assessed competence, participants in the experimental group reported substantial increases in their perceived ability to perform value stream analysis and design. Improvements ranged from 7.1 to 35.7%, with moderate to large effect sizes. Although not all p-values reached formal statistical significance, the consistent upward trends indicate that simulation game-based learning positively influenced participants' confidence and perceived mastery. This self-perception is a relevant factor in applied learning contexts, as it often correlates with increased motivation, willingness to apply knowledge, and long-term retention.

Finally, the user feedback confirmed that the game was well received by participants. Most respondents stated that the simulation improved their understanding of key concepts and supported their learning process. The game dynamics—particularly the integration of collaboration, competition, and immediate feedback—were seen as motivating and engaging. These subjective impressions are consistent with the measured improvements and underline the potential of the simulation game as a format of initial education and professional development for teaching complex methods like value stream design contexts.

Taken together, the results suggest that "North Star—The Value Stream Game" successfully addresses the complexity of value stream design as a structured, interactive, and socio-technical simulation game. The integration of LMMs, resource constraints, and decision consequences creates a rich setting that supports and guides knowledge transfer, practical application, and competence development. While the sample size limits the generalizability of the findings, the statistically and practically significant effects point toward the relevance of simulation game-based learning in engineering education.

Future work should aim to expand the sample size, explore variations in group composition and instructional integration, and investigate long-term learning retention. In addition, the potential of passive learning through observation, as demonstrated by the control group's results, offers a valuable direction for further research into hybrid learning settings.

4.7 Conclusion and Outlook

This chapter presented the application of a novel CRISP-DM-based gamification framework through the development of a socio-technical simulation game titled "North Star—The Value Stream Game", specifically designed to effectively support initial education and ongoing professional development in the field of value stream design. By integrating lean management, gamification, and LLMs this simulation game provides a structured, interactive, and immersive system. The approach empowers human improvement teams to bridge the gap between theoretical knowledge and practical application, systematically enhancing their competencies across defined levels of mastery—ranging from foundational understanding and practical application to mastery and even training proficiency.

The empirical evaluation demonstrated that the simulation game significantly supports measurable learning gains across several key dimensions after just one intervention. Participants in the experimental group achieved statistically significant improvements in knowledge acquisition ($p = 0.002$), practical skills development ($p < 0.001$), and self-assessed competence ($p = 0.038$), validating the high educational effectiveness of this approach. Notably, even passive observers in the control group benefited from socio-technical immersion, highlighting additional learning effects beyond active participation. Furthermore, user feedback confirmed high motivation, engagement, and perceived educational value.

A particular strength of the socio-technical simulation game lies in its ability to systematically translate complex domain knowledge into actionable learning experiences. The embedded AI-based digital systems actively support and guide human improvement teams as they navigate realistic constraints. Consequently, learners successfully progress through the various levels of mastery, acquiring and applying resource efficient thinking effectively in practice-oriented scenarios.

However, limitations such as the relatively small sample size and single-institution context must be acknowledged. These factors restrict the generalizability of results, highlighting the need for further studies involving broader groups of participants, longer evaluation periods, and more diverse instructional settings.

Overall, the presented socio-technical simulation game "North Star—The Value Stream Game" exemplifies the significant potential of integrating advanced methodologies and game-based learning with AI in both initial education and professional development. It provides a robust foundation for equipping human improvement teams with the competencies required to address the increasing complexity of value stream design, ultimately fostering resource efficiency and strategic thinking in a digitally empowered world.

References

1. Rother, M., Shook, J.: Sehen lernen, Mit Wertstromdesign die Wertschöpfung erhöhen und Verschwendung beseitigen. Lean Management Institut GmbH (Workbooks für Lean Management), Mülheim a.d. Ruhr (2011)
2. Erlach, K.: Wertstromdesign, Der Weg zur schlanken Fabrik. Springer-Verlag GmbH, Berlin (2010)
3. Engel, L., Kranhold, M.: Lean Management wird zum Green Management, Um Materialflusskosten erweiterte Wertstromanalyse. Dt. Ing.-Bl. **2021**, 1–6 (2021)
4. Meudt, T.: Wertstromanalyse 4.0 – Eine Methode zur integrierten Erfassung und Analyse von Material- und Informationsflüssen in Wertströmen. Dissertation, Technische Universität Darmstadt. Shaker Verlag, Düren (2020)
5. Meudt, T., Kaiser, J., Metternich, J., Spieckermann, S.: Wertstrommodellierung und-simulation im Zeichen von Digitalisierung und Industrie 4.0. ZWF **112**, 865–868 (2017)
6. Meudt, T., Metternich, J., Abele, E.: Value stream mapping 4.0, holistic examination of value stream and information logistics in production. CIRP Ann. Manuf. Technol. **66**, 413–416 (2017)
7. Zanker, C., Reisen, K.: Stabilitäts-und Flexibilitätsanforderungen an Produktionssysteme. In: Kötter, W., Schwarz-Kocher, M., Zanker, C. (eds.) Balanced GPS. Springer Fachmedien, Wiesbaden (2015)
8. Geisthardt, M., Engel, L.: A contribution to the development of sustainable target value streams with machine learning considering material flow costs. In: Marx Gómez, J., Nyambo, D.G., Sam, A.E. (eds.) Artificial Intelligence Tools and Applications in Embedded and Mobile Systems. Springer Nature, Berlin (2024)
9. Geisthardt, M., Engel, L., Döhr, J., Marx Gómez, J.: Towards digital solutions for value stream analysis and design: investigating enterprise perspective requirements. In: Herberger, D., Hübner, M. (eds.) Proceedings of the Conference on Production Systems and Logistics (CPSL), pp. 346–356. Hannover (2024)
10. Geisthardt, M., Engel, L., Meyer, L., Marx Gómez, J.: Towards digital solutions for value stream analysis and design: systematic literature review and market analysis. In: Herberger, D., Hübner, M. (eds.) Proceedings of the Conference on Production Systems and Logistics (CPSL), pp. 357–367. Hannover (2024)
11. Liker, J.: The Toyota Way. McGraw-Hill, New York (2004)
12. Balaji, V., Venkumar, P., Sabitha, M., Amuthaguka, D.: DVSMS: dynamic value stream mapping solution by applying IIoT. Sādhanā **45**, 1–13 (2020)
13. de Assis, R.F., de Santa-Eulalia, L.A., Ferreira, W.d.P., Armellini, F., Anholon, R., Rampasso, I.S., Santos, J.: Translating value stream maps into system dynamics models: a practical framework. Int. J. Adv. Manuf. Technol. **114**, 3537–3550 (2021)
14. Frick, N., Metternich, J.: The digital value stream twin. Systems **10**, 102 (2022)
15. Lu, Y., Liu, Z., Min, Q.: A digital twin-enabled value stream mapping approach for production process reengineering in SMEs. Int. J. Comput. Integr. Manuf. **34**, 764–782 (2021)
16. Frick, N., Urnauer, C., Metternich, J.: Echtzeitdaten für das Wertstrommanagement. ZWF **115**, 220–224 (2020)

17. Teriete, T., Böhm, M., Sai, B.K., Erlach, K., Bauernhansl, T.: Event-based framework for digitalization of value stream mapping. Procedia CIRP **107**, 481–486 (2022)
18. iFAKT GmbH: iFAKT website (2024). https://www.ifakt.de
19. IBM Deutschland GmbH: IBM process Mining (2024). https://www.ibm.com/products/process-mining
20. ONIQ GmbH: ONIQ (2024). https://oniq.com/de
21. Robert Bosch Manufacturing Solutions GmbH: NEXEED Industrial Application System(2024). https://www.bosch-connected-industry.com/de/de/portfolio/nexeed-industrial-application-system
22. Rother, M.: Toyota Kata: Managing People for Improvement, Adaptiveness and Superior Results. McGraw-Hill, New York (2009)
23. Geisthardt, M., Engel, L., Marx Gómez, J.: Simulation game based teaching of value stream analysis and design with Multimodal Large Language Models (MLLMs) assistance. In: Mathis, U., Ondrusch, N., Kilian, D., Redlein, A., Turowski, K., Weidner, S., Wittges, H. (eds.) Proceedings of the SAP Academic Community Conference 2024 (D-A-CH). Technische Universität München, Munich (2024)
24. Geisthardt, M., Engel, L., Schnegelberger, M.: Simulation game concept for AI-enhanced teaching of advanced value stream analysis and design. In: Herberger, D., Hübner, M. (eds.) Proceedings of the Conference on Production Systems and Logistics (CPSL). publish-Ing., Hannover (2023)
25. Geisthardt, M., Engel, L., Schnegelberger, M.: Towards gamification of advanced value stream analysis and design: a game-based learning concept. In: Wohlgemuth, V., Kranzlmüller, D., Höb, M. (eds.) LNI EnviroInfo 2023. Gesellschaft für Informatik e.V., Bonn (2023)
26. Huang, D., Yan, C., Li, Q., Peng, X.: From large language models to large multimodal models: a literature review. Appl. Sci. **14**, 5068 (2024)
27. Brocke, J., van der Aalst, W.M., Berente, N., van Dongen, B., Grisold, T., Kremser, W., Mendling, J., Pentland, B.T., Roeglinger, M., Rosemann, M.: Process science: the interdisciplinary study of socio-technical change. Process Sci. **2**, 1 (2024)
28. Kim, T.K.: T test as a parametric statistic. Korean J. Anesthesiol. **68**, 540–546 (2015)
29. Rasch, B., Friese, M., Hofmann, W., Naumann, E.: Quantitative Methoden 1. Einführung in die Statistik für Psychologen und Sozialwissenschaftler. Springer, Berlin (2014)
30. Ohno, T.: Das Toyota-Produktionssystem. Campus Verlag GmbH, Frankfurt/Main (2013)
31. Deutsches Institut für Normung e.V.: DIN EN ISO 14051:2011–12, Umweltmanagement – Materialflusskostenrechnung – Allgemeine Rahmenbedingungen (ISO 14051:2011). Beuth Verlag, Berlin (2011)
32. Trist, E.L., Bamforth, K.W.: Some social and psychological consequences of the longwall method of coal-getting: an examination of the psychological situation and defences of a work group in relation to the social structure and technological content of the work system. Hum. Relat. **4**, 3–38 (1951)
33. Baxter, G., Sommerville, I.: Socio-technical systems: from design methods to systems engineering. Interact. Comput. **23**, 4–17 (2011)
34. Emery, F.E., Trist, E.L.: Socio-technical systems. In: Churchman, C.W., Verhulst, M. (eds.) Management Science: Models and Techniques, vol. 2, pp. 83–97. Pergamon Press, London (1960)
35. Sarker, S., Chatterjee, S., Xiao, X., Elbanna, A.: The sociotechnical axis of cohesion for the IS discipline. MIS Q. **43**, 695-A695 (2019)
36. Bednar, P.M., Welch, C.: Socio-technical perspectives on smart working: creating meaningful and sustainable systems. Inf. Syst. Front. **22**, 281–298 (2020)
37. Yu, D., Glazieva, A.: Game based learning with artificial intelligence and immersive technologies: an overview. CEUR Workshop Proc. **3077**, 146–159 (2022)
38. Ma, J., Zhang, Y., Bin, H., Wang, K., Liu, J., Gao, H.: The development of students' computational thinking practices in AI course using the game-based learning: A case study. In: 2022 International Symposium on Educational Technology (ISET), pp. 273–277. IEEE, Hong Kong (2022)

39. Samiha, M., Akram, B., Barnes, T., Price, T.: Adaptive immediate feedback for block-based programming: design and evaluation. IEEE Trans. Learn. Technol. **15**, 406–420 (2022)
40. Élise, L., Monterrat, B., Desmarais, B., George, S.: Adaptive gamification for learning environments. IEEE Trans. Learn. Technol. **12**, 16–28 (2019)
41. Troussas, C., Krouska, A., Virvou, M.K.: Using a multi module model for learning analytics to predict learners' cognitive states and provide tailored learning pathways and assessment. In: Virvou, M., Alepis, E., Tsihrintzis, G.A., Jain, L.C. (eds.) Machine Learning Paradigms. Studies in Computational Intelligence, vol. 940, pp. 85–108. Springer, Cham (2020)
42. Brown, T.B., et al.: Language models are few-shot learners (2020). https://arxiv.org/abs/2005.14165
43. Vaswani, A., et al.: Attention is all you need (2017). https://arxiv.org/abs/1706.03762
44. Koehn, P.: Neural machine translation. Cambridge University Press, Cambridge (2020)
45. Jurafsky, D., Martin, J.H.: Speech and language processing: an introduction to natural language processing, computational linguistics, and speech recognition, 3rd edn. (2021). https://web.stanford.edu/~jurafsky/slp3/. Accessed 18 July 2025
46. Bommasani, R., et al.: On the opportunities and risks of foundation models (2021). https://arxiv.org/abs/2108.07258
47. Baltrušaitis, T., Ahuja, C., Morency, L.-P.: Multimodal machine learning: a survey and taxonomy. IEEE Trans. Pattern Anal. Mach. Intell. **41**, 423–443 (2018)
48. Lenarduzzi, V., Taibi, D.: MVP explained: a systematic mapping study on the definitions of minimal viable product. In: 42nd Euromicro Conference on Software Engineering and Advanced Applications (SEAA), pp. 112–119. IEEE (2016)
49. Ries, E.: Minimum viable product: a guide (2009). https://www.startuplessonslearned.com/2009/08/minimum-viable-product-guide.html. Accessed 18 July 2025
50. Wirth, R., Hipp, J.: CRISP-DM: towards a standard process model for data mining. In: 4th International Conference on the Practical Applications of Knowledge Discovery and Data Mining (PADD), Manchester, UK, April 2000, pp. 29–39 (2000)
51. Werbach, K., Hunter, D.: For the win: the power of gamification and game thinking in business, education, government, and social impact, rev. and updated edn. University of Pennsylvania Press, Philadelphia (2020)
52. Johnson, C.N.: The benefits of PDCA. Qual. Prog. **49**, 45 (2016)

Chapter 5
A Data-Driven Clustering Approach for Assessing Service Performance of Brand Chains' Branches in the Food Service Industry

Musa Yılmaz, Ayşenur Karakaya, Ali Pişirgen, and Serhat Peker

Abstract This study proposes a data-driven clustering methodology to evaluate the service performance of foodservice brand chain branches in Türkiye by analyzing customer-generated content from Google Reviews. The research identifies key performance indicators, total and recent review frequency and ratings and applies k-means clustering to segment 735 branches from five major chains into distinct performance profiles. Cross-analysis with contextual variables such as sector, location, and urbanization level provides insights into the drivers of customer satisfaction. Results reveal significant performance variations among branches, highlighting clusters of high satisfaction and clusters with persistent issues, particularly among mall-based and fast-food locations. The findings underscore the utility of online review data for performance benchmarking and strategic improvement in the foodservice sector. This clustering approach offers both researchers and practitioners a practical tool for assessing service quality and enhancing customer satisfaction.

Keywords Service performance · Food service industry · Google reviews · Clustering · Customer satisfaction · Türkiye

5.1 Introduction

The food service industry is rapidly evolving, driven by economic growth and a thriving tourism sector. According to a recent report by Mordor Intelligence, Türkiye's foodservice market is projected to reach $13.88 billion in 2024 and grow to

M. Yılmaz · A. Karakaya · S. Peker (✉)
Management Information Systems, Izmir Bakircay University, İzmir, Turkey
e-mail: serhat.peker@bakircay.edu.tr

A. Pişirgen
Department of Management and Organization, Karamanoğlu Mehmetbey University, Karaman, Türkiye

M. Mora et al. (eds.), *Engineering and Management of Data Science, Analytics, and AI/ML Projects*, Intelligent Systems Reference Library 282,
https://doi.org/10.1007/978-3-032-06889-7_5

$24.10 billion by 2029, reflecting a compound annual growth rate (CAGR) of 11.66% over this period [1]. This expansion highlights the industry's dynamic nature and its continuous adaptation to shifting customer demands, making demand analysis crucial for businesses striving to remain competitive [2].

In an increasingly competitive service landscape, understanding and improving service performance is essential for maintaining customer satisfaction and ensuring long—term success [3]. For chain restaurants, systematically evaluating the performance of all branches in terms of customer satisfaction provides a comprehensive view of their service quality and helps identify areas for improvement. In today's digital era, online reviews, particularly those on platforms like Google Reviews, play a pivotal role in shaping customer decisions [4]. The vast amount of feedback available on these platforms allows businesses to assess customer satisfaction, enhance their reputation, and refine their overall customer experience.

Google Reviews has become an indispensable tool for both consumers and businesses in the food and beverage industry [5]. Customers increasingly rely on online reviews to guide their dining choices, often favoring establishments with higher ratings and positive feedback. For brand chains, Google Reviews offer invaluable insights into customer preferences, satisfaction drivers, and areas of dissatisfaction [6]. By systematically analyzing these reviews, businesses can gain a deeper understanding of their branches' performance, implement necessary improvements, and ultimately foster higher service quality, increased customer satisfaction, and greater customer loyalty [7].

This study introduces a data-driven clustering approach to assess the service performance of brand chain branches by leveraging extensive customer feedback from Google Reviews. The proposed method incorporates key factors such as the number of reviews and average review scores and employs the k-means clustering algorithm to identify distinct branch profiles based on service quality. Additionally, a cross-analysis of the generated clusters is conducted, considering descriptive features such as business sector, branch location, and the urbanization level of the geographical area. The methodology is applied to five major brand chains in Türkiye, covering restaurants, patisseries and bakeries, and coffeehouses. The resulting clusters are analyzed in depth to identify the characteristics that differentiate high-performing branches from those requiring improvement.

The findings of this study highlight significant variations in customer satisfaction across branches, offering actionable insights for brand chains to enhance service quality, improve customer satisfaction, and optimize operational strategies. By implementing this clustering approach, businesses in the foodservice industry can gain a comprehensive understanding of branch performance, enabling targeted interventions, data-driven decision-making, and strategic improvements for sustained growth and competitiveness.

5.2 Literature Review

5.2.1 Online Reviews and Ratings as Metrics for Service Performance and Customer Satisfaction

Online consumer reviews have become a crucial barometer for service performance and customer satisfaction in modern business evaluation. Across various service industries, studies show that the content and ratings of online reviews closely reflect customers' actual experiences and perceptions. In the hospitality sector, for example, researchers have linked online feedback to tangible performance outcomes. In [8], hotel reviews and management responses in a panel study of 843 hotels are analysed, finding that eWOM (electronic word-of-mouth) metrics—such as review ratings, volume, and variance—are significantly associated with hotel performance. Similarly, in [9] text mining is applied to Booking.com hotel reviews to identify key satisfaction and dissatisfaction features, demonstrating that review text can reveal what aspects of service (rooms, cleanliness, etc.) drive customer satisfaction. In the context of tourism lodging, in [10] online reviews for Spanish rural accommodations are examined and observed that effective management of online word-of-mouth correlates with improved business performance, underscoring the influence of customer feedback on firm success. These works affirm that online reviews serve as a valuable proxy for measuring service quality and satisfaction levels, often complementing or even substituting traditional survey-based assessments.

In the foodservice industry, online reviews are equally pivotal for evaluating and comparing service performance. Large-scale analyses of restaurant reviews have revealed insights into customer satisfaction trends. Google review data (over 82,000 reviews) to compare small fast-food businesses with national chain restaurants are used, uncovering differences in customer preferences and satisfaction between independent eateries and big-brand outlets [11]. A follow-up study by [11] similarly analysed about 80,000 Google reviews from quick-service restaurants, confirming that local restaurants and chains exhibit distinct patterns in review content and ratings. Such findings illustrate how online feedback can highlight strengths and weaknesses for different business models (e.g. consistent quality in chains vs. personalized service in local venues). The influence of online ratings on consumer behaviour is well documented by [12] presenting a conjoint analysis that review features like food quality, service quality, and atmosphere play a critical role in restaurant selection decisions, indicating that customers heavily weigh these aspects when choosing where to dine. Moreover, the credibility of reviewers and the trustworthiness of their opinions can modulate business outcomes. In [13], the perceived trustworthiness of Yelp reviewers (modelled via a logistic regression on reviewer and review features) impacts business patronage, highlighting that not only the content of reviews but also the reputation of reviewers can shape consumer perceptions. Together, these studies emphasize that online review platforms (e.g. Google, Yelp, TripAdvisor) are indispensable resources for gauging customer satisfaction and guiding service improvements in the restaurant industry.

The importance of online reviews extends to transportation and other service domains as well. In the aviation sector, over half a million TripAdvisor airline reviews to extract service quality indicators (using topic modelling and sentiment analysis), improving predictive models of passenger satisfaction by [14]. Their work demonstrated that unstructured review data can reveal factors like seat comfort and staff behaviour that strongly influence overall satisfaction, thereby informing airline service enhancements. In [15], sentiment analysis and topic modelling are employed on Google Reviews to assess airport service quality, showcasing that crowdsourced online feedback correlates well with official customer satisfaction surveys. In fact, certain themes from Google review text (e.g. security, cleanliness) aligned with established ASQ (Airport Service Quality) dimensions, suggesting that online reviews can complement traditional survey methods in evaluating service performance [15]. Beyond travel, researchers have even compared online review metrics with formal service quality inspections to validate the informational value of consumer feedback. In [16], authors examined whether restaurant food safety inspections align with public perceptions on Yelp and Google, finding notable associations between poor inspection results and lower online ratings. Likewise, in [17], 50,000+ TripAdvisor restaurant reviews were analysed alongside official health inspection scores in New York, reporting significant correlations that imply online review sentiment can serve as an early indicator of potential service issues (in this case, food safety problems). Such cross-comparative studies reinforce that onlinc reviews and ratings are not only shaping customer decisions but are increasingly recognized by researchers as reliable data points for measuring service quality and customer satisfaction across diverse sectors.

5.2.2 Data Mining and Machine Learning Techniques for Review Analysis

Given the rich, unstructured nature of online reviews, researchers have adopted a wide range of data mining and machine learning techniques to extract actionable insights. A central approach in many studies is sentiment analysis, which involves computationally determining whether a review's tone is positive, negative, or neutral. Sentiment analysis allows quantification of customer satisfaction from textual data at scale. Numerous studies have applied this technique to online reviews to gauge service performance. For instance, in [15], sentiment analysis is used to convert Google review text into sentiment scores, which they then compared with star ratings and survey results to evaluate airport services. In [18], authors focused on capturing "emotionally colored" information in customer reviews, using sentiment analysis and natural language processing (NLP) to classify the emotional tone of service feedback. By accounting for emotions expressed in text, Yusupova's analysis provided a deeper understanding of customer attitudes to aid in decision-making. In the realm of hospitality, sentiment analysis on mobile app service reviews is performed and then a gap analysis is conducted, identifying gaps between customer expectations and actual perceptions for a navigation service [19]. This approach of combining sentiment mining with a service feature hierarchy enabled a diagnostic view of where a service was underperforming. To improve the accuracy of sentiment extraction, researchers have also experimented with different algorithms: the performance of Support Vector Machines (SVM) and Random Forest classifiers were compared for labeling tourism reviews as positive or negative, finding that machine learning models can reliably categorize sentiment to inform strategic decisions for tourism services [20]. Similarly, in [21], authors applied an ensemble of algorithms (Logistic Regression, SVM, Naïve Bayes, Random Forest, and XGBoost) on airport review data after converting text to features (using TF-IDF), in order to predict sentiment classes. Such efforts illustrate the prevalent use of sentiment analysis and supervised learning to transform unstructured review text into quantitative measures of satisfaction, which can be tracked and analyzed over time. Furthermore, this study introduces IVADER, an enhanced VADER-based sentiment analysis model that builds domain-specific sentiment lexicons using polarity and entropy calculations to classify customer reviews more accurately across multiple domains. IVADER achieves high accuracy (98.64%) and outperforms traditional models in terms of precision, recall, specificity, and training efficiency, demonstrating strong potential for multi-domain sentiment evaluation [22].

Beyond sentiment polarity, topic modeling and other NLP techniques have been widely used to uncover prevalent themes in customer feedback. These unsupervised methods automatically discover the topics or aspects that customers discuss in their reviews, providing insight into what service features matter most to consumers. A common technique is Latent Dirichlet Allocation (LDA), which has been utilized in several studies. LDA was applied to hundreds of thousands of Booking.com hotel reviews from Bogotá and Madrid, identifying key aspects of hospitality service quality mentioned by guests [23]. This analysis, combined with word embeddings (FastText) and dimensionality reduction, helped pinpoint which service elements (such as room cleanliness, staff friendliness, or location convenience) were most salient in positive versus negative reviews. Topic modeling was taken a step further by conducting a cross-sector analysis using E-LDA (an extended LDA algorithm) on an enormous dataset of over 32 million reviews spanning industries from healthcare to online gaming [24]. By extracting bigram and trigram topics and tying them to quality management frameworks (like SERVQUAL dimensions), Rajendran's meta-analysis was able to highlight common service quality pain points and best practices across different sectors. In [25], authors employed a case study on Singapore hotels demonstrates that proposed framework to evaluate the importance of hotel service features on customer satisfaction using interpretable machine learning (IML) offers more accurate, interpretable, and consistent importance estimates than traditional methods, aiding service improvement efforts. Other researchers have used alternative topic modeling approaches; for example, in [14], authors employed a Structural Topic Model (STM) on airline reviews, which allowed them to incorporate metadata (like star rating and flight characteristics) as covariates in the topic extraction process. This yielded a nuanced understanding of how certain topics (e.g., "seat comfort" or "customer service") were associated with higher or lower satisfaction levels. Additionally, some studies use specialized text analytics tools: Leximancer, as a concept mapping software, was utilized to perform a thematic analysis of 3,951 TripAdvisor reviews for a unique case (restaurants run in prisons) [26]. This qualitative-driven tool automatically clustered concepts from the reviews, revealing prominent themes in customers' dining experiences (such as the "food" quality and the novelty of the "prison" setting) that traditional metrics might overlook. Through topic modeling and NLP, these works demonstrate how unstructured reviews can be distilled into meaningful categories and themes, enriching the evaluation of service performance with customers' own words.

In parallel, many studies have developed predictive models and classification frameworks to leverage review data for specific outcomes. Rather than only describing sentiments or topics, these approaches use machine learning to predict or classify variables of interest based on review-derived features. One stream of research focuses on predicting review helpfulness or trust, which has implications for platform design and business reputation. In [27], authors built machine learning models (using features such as review polarity, subjectivity, and readability scores) to predict the helpfulness of Amazon product reviews. By analyzing over 622,000 reviews, their model could identify which reviews were likely to be voted as "helpful" by other consumers, providing insights into the characteristics of impactful customer feedback. In [28], By integrating machine learning with traditional service quality frameworks, the study offers practical and theoretical insights to improve OFD service operations and consumer satisfaction. In [13], as noted earlier, he developed a predictive model for reviewer trustworthiness on Yelp: using logistic regression, the study classified reviewers as trustworthy or not based on their history and features, linking this to how much reviews influence business patronage. Other studies integrate review analytics with external data to inform strategic decisions. An integrated framework for hotel chains, where sentiment analysis of Hilton hotel reviews was combined with a convolution tree kernel classifier was presented to perform aspect-based sentiment classification [29]. This framework also incorporated Google Trends data to visualize interest over time, creating a comprehensive dashboard of customer sentiment and brand reputation. This kind of integration shows the power of machine learning to not only parse reviews but also connect them to broader business intelligence indicators. In the public sector domain, an XGBoost machine learning classifier was utilized on Google Maps reviews to support participatory urban planning [30]. By classifying and interpreting thousands of park review comments in Istanbul, their approach identified key factors (like cleanliness, facilities, and accessibility) that city planners could address to improve public spaces, illustrating an innovative use of review data for service design. Furthermore, some researchers have blended textual analysis with advanced statistical modeling. In [11], for instance, sentiment text analysis with multilevel multinomial modeling was paired to compare customer satisfaction between local vs. chain restaurants, effectively bridging unstructured and structured data analysis. This combination allowed controlling context (local vs. chain) while leveraging the rich information in review text, yielding more robust conclusions about service differences. Overall, the literature showcases a broad arsenal of data mining techniques, ranging from NLP methods like sentiment analysis and topic modeling to supervised learning algorithms and hybrid statistical models, all aimed at extracting insights from unstructured online reviews. By applying these techniques, prior studies have been able to quantify customer satisfaction, uncover the topics that matter most to consumers, and even predict future customer behaviors or outcomes based on review patterns.

The present study introduces a k-means clustering approach to review-based service evaluation, addressing the lack of clustering applications in the literature. By analyzing a large corpus of Google Reviews from multiple branches of major food and beverage chains, this study demonstrates how unsupervised clustering can be used to evaluate service performance in a novel way. We use review metrics (such as the volume of reviews and average ratings for each branch) as input features for clustering, thereby grouping service outlets into distinct performance profiles based on customer feedback patterns. Through this methodology, we uncover natural groupings of branches—for instance, clusters of consistently high-performing outlets versus clusters of under-performing ones—without relying on predetermined labels. This clustering-based evaluation provides a fresh perspective that complements existing techniques like sentiment analysis and regression modeling. By identifying performance clusters and then examining their characteristics (e.g. do low-performing branches share common issues in customer comments), we offer actionable insights that would be difficult to obtain from sentiment scores or topic frequencies alone. In summary, this work positions itself as a direct response to the identified methodological gap: it demonstrates the value of clustering in making sense of unstructured review data for service quality assessment, thereby extending the analytical toolkit for online review research beyond the more commonly used predictive and text-mining methods. Through this contribution, the study not only enriches the academic literature but also provides practitioners with a practical approach to benchmark and improve service performance using the wealth of information contained in online customer reviews.

5.3 Methodology

This study adopts a data-driven approach to evaluate the service performance of foodservice brand chain branches by leveraging online customer feedback. Google Maps reviews, a widely accessible and influential source of consumer opinion, serve as the primary dataset for this research. To systematically analyze the variations in service quality across multiple branches, the study defines key performance features, collects and prepares relevant data, and applies unsupervised machine learning through cluster analysis. By employing the k-means algorithm, the research aims to group branches based on similarities in review frequency and customer satisfaction ratings, both historically and in recent periods. This methodology enables the identification of distinct performance profiles, offering valuable insights for benchmarking and strategic decision-making within the competitive food service industry.

The following sections detail the study's feature selection, data collection procedures, data preparation techniques, and clustering methodology.

5.3.1 Features

Four key performance metrics as features were defined for each branch to quantify service performance and customer feedback:

- **Frequency**: The total number of customer reviews the branch has received since its listing on Google. This serves as a cumulative indicator of customer engagement over time.
- **Rating**: The overall average star rating (out of 5) of all reviews for the branch. This metric reflects the aggregate customer satisfaction level throughout the branch's history.
- **Recent Frequency**: The number of reviews received in the last month. This captures recent customer engagement, indicating how active customers have been in providing feedback in the short term.
- **Recent Rating**: The average star rating of reviews from last month. This provides insight into the branch's recent service quality, highlighting current customer satisfaction trends.

These features offer both long-term and short-term perspectives on performance, allowing us to assess overall standing (Frequency and Rating) as well as recent developments or changes in customer feedback (Recent Frequency and Recent Rating).

5.3.2 Data Collection

The data used in this study were manually collected by the researchers from customer reviews available on the Google Maps platform. This manual approach ensured the accuracy and relevance of the extracted information for each branch. To minimize temporal discrepancies and enhance comparability across branches, the data gathering process was completed within a short timeframe of approximately three days. This strategy enabled the researchers to capture a synchronized snapshot of branch performance, reducing the influence of fluctuations in daily review activity. The consistent and careful manual collection process provided a reliable dataset for evaluating and comparing the service performance of foodservice brand chain branches.

5.3.3 Data Preparation

After data collection, the dataset was carefully cleaned and prepared to ensure reliability. All branches' records were verified, and any entries with missing values or inconsistencies were removed or corrected. This cleaning step ensured that only complete and accurate data were used in the analysis, eliminating potential biases or errors due to incomplete information.

Each branch's data was then organized to include not only the performance metrics above, but also contextual descriptors to aid interpretation. In particular, the dataset recorded the branch's brand affiliation, category (e.g. restaurant, patisserie, coffee-house), geographic location (province and district), the urbanization level of its location (urban or rural), and whether the branch is located in a shopping mall. While these contextual features were not used as inputs for clustering, they were retained for post-clustering analysis to help explain and compare cluster characteristics (for example, seeing if certain clusters correspond to particular regions or business types).

Prior to applying the clustering algorithm, all numeric performance metrics were standardized using Z-score normalization. Each metric was mean-centered and scaled to unit variance (zero mean and standard deviation of one). This standardization ensured that metrics with larger numerical ranges (such as review counts) did not unduly dominate those with smaller ranges (like average ratings) in the clustering process. By putting Frequency, Rating, Recent Frequency, and Recent Rating on a comparable scale, we facilitated a fair and balanced clustering analysis.

5.3.4 Cluster Analysis

The core analytical technique employed to group the branches was the cluster analysis, with the k-means algorithm selected as the clustering method due to its effectiveness in partitioning data into distinct, non-overlapping groups based on similarity. K-means is an unsupervised machine learning method that assigns data points (in this case, branch performance profiles) into k distinct clusters. It operates iteratively, starting from an initial random seeding of k cluster centroids and then repeatedly updating cluster memberships and centroids. In each iteration, branches are assigned to the nearest centroid (based on Euclidean distance in the multidimensional feature space of the four-performance metrics), and then the centroid positions are recalculated as the mean of all points in the cluster. This process continues until convergence, minimizing the within-cluster sum of squares (i.e., the variance) for each cluster. The result is a division of branches into clusters such that branches within the same cluster have very similar performance metric profiles, while those in different clusters are as distinct as possible.

Determining an appropriate number of clusters (k) is crucial for meaningful results. We utilized the Elbow method to guide the selection of k [31]. In the Elbow method, the clustering algorithm is run for a range of different k values, and for each k

we calculate the total within-cluster variance (often measured by the sum of squared distances of each point from its cluster centroid). We then plot this variance against the number of clusters. Typically, as k increases, the within-cluster variance decreases (clusters become smaller and tighter). The Elbow method identifies the point in this plot where increasing k further yields only marginal improvements—this point of inflection resembles an "elbow." That elbow point is considered an indicator of a suitable number of clusters beyond which additional clusters would not significantly improve the model.

5.4 Case Study

5.4.1 Sample and Descriptive Statistics

This study focuses on the rapidly expanding foodservice industry in Türkiye, which has been experiencing significant growth due to economic development and increasing consumer demand. To ensure comprehensive coverage across the sector, the sample includes the branches of five leading brand chains, each representing a different segment of the industry: patisserie and bakery, coffeehouse, fast-food restaurant, and fast casual restaurant.

The selected brands are among the most prominent and widely recognized in the Turkish foodservice market. All available branches of these chains were included in the analysis to capture a holistic view of their service performance. Specifically, in Table 5.1, the sample consists of the following:

This diverse and sectorally representative sample of 735 branches in total provides a robust foundation for clustering analysis.

Table 5.2 summarizes the distribution of the four primary performance metrics used in the clustering analysis, including total and recent review frequency and that of rating. Values represent the maximum, minimum, mean, and standard deviation across all branches in the dataset. The analysis revealed considerable variation across branches. To detail, the total number of reviews (Frequency) per branch ranged from as low as 20 to as high as 5,747, indicating a wide disparity in customer exposure and engagement among branches. Similarly, average ratings spanned from 1.8 to 4.7

Table 5.1 Number of branches by brand

Brand name	Number of branches
Bolulu Hasan Usta (patisserie & bakery)	42
Özsüt (patisserie & bakery)	91
Kahve Dünyası (coffeehouse)	103
McDonald's (fast-food restaurant)	212
Tavuk Dünyası (fast casual restaurant)	287

Table 5.2 Descriptive statistics of performance metrics

	Maximum	Minimum	Mean	Standard deviation
Frequency	5,747.00	20.0	491.69	618.63
Rating	4.7	1.8	3.64	0.46
Recent frequency	101.00	1.00	11.74	12.63
Recent rating	5.00	1.00	3.24	1.09

out of 5, with an overall mean around 3.64, suggesting diverse levels of customer satisfaction.

5.4.2 Determination of Optimum Number of Clusters

Based on the Elbow plot (see Fig. 5.1) for our dataset, we observed a noticeable bend at $k = 3$, but the within-cluster sum of squares (WCSS) continued to decrease significantly until $k = 7$. Specifically, the WCSS dropped by more than 500 units between $k = 3$ and $k = 7$, after which the rate of decrease became marginal. This suggests that $k = 7$ provides a better balance between cluster cohesion and explanatory power. Therefore, we selected $k = 7$ as the optimal number of clusters for the final clustering solution. The k-means algorithm was applied to the standardized data with $k = 7$, producing seven clusters of branches with similar service performance characteristics. Each branch was thus assigned to one of seven clusters, and these cluster groupings form the basis for further analysis. In the subsequent results, we examine the characteristics of these clusters to interpret the distinct performance profiles and to identify the factors differentiating high-performing branches from those in need of improvement.

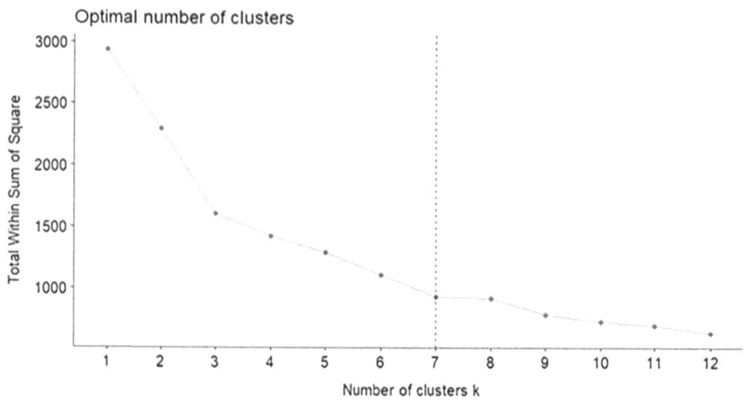

Fig. 5.1 Elbow method for determining the optimal number of clusters

5.4.3 Results of Cluster Analysis

The k-means clustering analysis segmented the foodservice branches into seven distinct clusters, each representing a unique performance profile based on Google Reviews data. As shown in Table 5.3, these clusters vary considerably in both the volume of customer feedback (review frequency) and satisfaction levels (average ratings). The optimal seven-cluster solution was determined using the elbow method (after evaluating several possible k values) to capture nuanced differences in branch performance. This segmentation reveals a spectrum of performance outcomes: from consistently high-performing clusters with strong customer satisfaction to clusters with significant performance issues.

In the following, we describe the characteristics of each cluster's performance and then compare their branch features to explain these differences.

High-Performing Clusters (Clusters 5 and 6): Clusters 5 and 6 demonstrate the strongest performance. Cluster 6 is the largest segment with 221 branches and boasts the highest overall satisfaction, with an average Rating of 4.06 (out of 5) and a Recent Rating of 4.15 (Table 5.3). This indicates not only high customer satisfaction but also a consistent or slightly improving trend in recent feedback. Cluster 6's branches receive a moderate number of reviews on average (about 321.58 total reviews per branch), suggesting they maintain high ratings even with substantial customer feedback. Similarly, Cluster 5 (24 branches) shows above-average performance: an overall Rating of 3.81 rising to a Recent Rating of 4.07, coupled with a surge in Recent Frequency (about 57 new reviews in the last period per branch). The improving trend in Cluster 5's ratings, alongside increased review counts, indicates that these branches have recently enhanced their service quality or customer experience, thereby boosting customer satisfaction and engagement. In summary, the high-performing clusters are characterized by strong ratings and upward or stable recent performance, marking them as benchmarks for successful service delivery.

Moderate-Performing Clusters (Clusters 1 and 2): Clusters 1 and 2 represent branches with average to slightly above-average performance. Cluster 1 is a large group of 188 branches that have a moderate review count (around 317 reviews on

Table 5.3 Cluster-wise results of branch performance metrics

Cluster	N	Frequency	Rating	Recent frequency	Recent rating
1	188	317.11	3.48	9.79	3.50
2	92	1,460.04	3.76	19.33	3.41
3	6	4,628.83	3.68	44.33	3.33
4	131	346.23	3.58	5.76	1.87
5	24	526.50	3.81	57.29	4.07
6	221	321.58	4.06	9.46	4.15
7	73	145.55	2.69	7.25	1.87
Average		**491.69**	**3.64**	**11.74**	**3.24**

average) and an overall Rating of 3.48 (Table 5.3). The Recent Rating for Cluster 1 remains steady at 3.50, indicating a stable performance that is slightly below the overall average. These branches are neither excelling nor failing dramatically; rather, they maintain an adequate level of customer satisfaction with room for improvement. Cluster 2 (92 branches) shows somewhat stronger baseline performance, with a higher average Rating of 3.76 and a very high volume of reviews (approximately 1,460 per branch on average). However, Cluster 2's Recent Rating has dipped to 3.41, down from its historical average. This decline, as shown in Table 5.3, suggests that despite strong customer engagement and past success, there may be emerging issues affecting customer satisfaction in Cluster 2 branches. In both clusters 1 and 2, performance is around the middle of the spectrum: generally acceptable ratings but not reaching the top-tier satisfaction levels, and in the case of Cluster 2, caution is warranted due to recent downward trends in reviews.

Low-Performing Clusters (Clusters 3, 4, and 7): The analysis identified three clusters that clearly lag in performance and warrant immediate attention. Cluster 7 exhibits the most pronounced underperformance. Comprising 73 branches, it has the lowest average Rating at just 2.69, with a critically low Recent Rating of 1.87 (Table 5.3). This cluster also has the fewest total reviews (around 146 on average), implying limited engagement that is largely negative, an indication of serious and persistent service quality problems. Cluster 4 (131 branches) likewise shows concerning results: its average Rating is 3.58, but recent customer feedback has plummeted (Recent Rating 1.87, identical to Cluster 7's recent score). Notably, Cluster 4's Recent Frequency of reviews is very low (only about 5.8 new reviews per branch), which, combined with the poor recent ratings, suggests that a small number of recent incidents or service lapses sharply drove down customer satisfaction. Meanwhile, Cluster 3 is an outlier in size and behavior, it contains only 6 branches, yet these branches have an extremely high review volume (over 4,600 total reviews on average, the highest of all clusters). Despite their popularity and reach, Cluster 3's average Rating is a moderate 3.68, and its Recent Rating has declined to 3.33. This pattern indicates that even highly frequented branches (likely flagship or very prominent outlets) can suffer from diminishing service quality, perhaps due to challenges in handling the high customer traffic or maintaining consistency. In summary, Clusters 3, 4, and 7 are characterized by subpar customer satisfaction. Cluster 7 and 4, in particular, stand out for their severe recent decline in service quality, while Cluster 3 signals that volume alone does not guarantee high ratings and may indeed strain service performance.

Overall, the cluster-wise performance metrics in Table 5.3 reveal a clear contrast between the high-satisfaction clusters (5 and 6) and the underperforming ones (especially 4 and 7), with clusters 1 and 2 in between. This stratification underscores the value of clustering: it allows us to pinpoint which groups of branches are excelling and which are struggling, providing a foundation for deeper investigation into the reasons behind these outcomes.

5.4.4 Results of Cross Analysis

To better understand the performance differences, we examined the distribution of branch characteristics for each cluster. Table 5.4 presents these characteristics in terms of location and business context: specifically, the proportion of branches located in shopping malls, the breakdown of each cluster by business sector (coffeehouse, fast-casual restaurant, fast-food restaurant, patisserie/bakery), and the percentage of branches in urban vs. rural areas.

Several notable patterns emerge from this cross-cluster comparison of features:

Location (Shopping Mall vs. Standalone): According to Table 5.4, the clusters differ in whether their branches tend to be situated inside shopping malls or in standalone locations. Notably, two of the low-performing clusters have the highest presence in shopping malls: Cluster 7 has about 80% of its branches located in shopping malls (the highest among all clusters), and Cluster 4 has 65% of its branches in malls. In contrast, the high-performing Cluster 5 has a much smaller mall presence (only 37.5% of its branches are in malls, meaning the majority are standalone locations), and Cluster 6 is roughly balanced (45% in malls vs. 55% outside). Clusters 2 and 3 also have low proportions of mall-based branches (19 and 17%, respectively, with the vast majority of those clusters' outlets located on high streets or other non-mall locations). This suggests that being located in a shopping mall is not inherently associated with better performance—in fact, our results show the opposite for some clusters. The two worst-rated clusters (4 and 7) are heavily mall-based, which could imply that certain mall environments or the conditions of operating in malls (such as higher foot traffic, competition, or perhaps reliance on captive mall audiences) do not automatically translate to high customer satisfaction. Branches in these clusters might face unique operational challenges or competitive pressures within mall settings that contribute to poorer service outcomes. On the other hand, branches

Table 5.4 Distribution of branch characteristics by cluster

Cluster	Inside shopping mall (%)		Sector distribution (%)				Rate of urban rural (%)	
	Yes	No	Coffeehouse	Fast casual restaurant	Fast-food restaurant	Patisserie and bakery	Urban	Rural
1	52.00	48.00	10.00	40.00	35.00	15.00	87.00	13.00
2	19.00	81.00	19.00	8.00	62.00	11.00	82.00	18.00
3	17.00	83.00	0.00	17.00	83.00	0.00	83.00	17.00
4	65.00	35.00	5.00	46.00	39.00	10.00	88.00	12.00
5	37.50	62.50	4.50	75.00	12.50	8.00	87.50	12.50
6	45.00	55.00	23.00	41.00	10.00	26.00	81.00	19.00
7	80.00	20.00	8.00	45.00	23.00	24.00	90.50	9.50
Average	**49.65**	**50.35**	**14.00**	**39.00**	**28.80**	**18.20**	**84.80**	**15.20**

operating outside of malls, as is common in Cluster 5 and much of Cluster 6, can achieve excellent performance, indicating that standalone venues are fully capable of delivering high customer satisfaction when managed effectively.

Urban vs. Rural Setting: The majority of branches across all clusters are in urban areas, reflecting the focus on major brand chains which tend to have an urban presence. However, there are slight differences in the degree of urban vs. rural representation that correlate with performance. Cluster 7, for instance, not only has the highest mall-based share but also the most urban-centric footprint, with about 90.5% of its branches in urban locations (Table 5.4). Yet this cluster performs the worst, underlining that an urban location alone does not guarantee success. In fact, the best-performing clusters did not have a significantly higher urban percentage than others: Cluster 6's branches are 81% urban (the lowest urban proportion among clusters, though still a large majority), and Cluster 5's are 87.5% urban, both slightly lower or comparable to the average urban rate (~85% across all clusters). Meanwhile, Cluster 6 includes the highest proportion of rural branches (19% rural, according to Table 5.4) among the clusters, and these branches still maintain high performance. This observation is insightful: it indicates that rural branches, which might be assumed to have lower resources or customer expectations, can nonetheless achieve excellent satisfaction levels (as seen in Cluster 6) when service quality is consistently high. In summary, all clusters are predominantly urban, but the presence of a few more rural outlets in a cluster is not associated with lower performance—if anything, Cluster 6 shows that even outside the big cities, branches can thrive. The extreme urban concentration of Cluster 7 does not help its performance, suggesting that factors beyond just urban demographics are influencing these outcomes (likely internal service issues, as discussed earlier).

Business Sector Composition: Each cluster's performance profile also appears to be linked with the type of foodservice business (sector) that predominates in that cluster (Table 5.4). For example, fast-food restaurants form the bulk of Clusters 2 and 3: Cluster 2 is about 62% fast-food outlets, and Cluster 3 is even more extreme at 83% fast-food. These clusters had only moderate to below-average ratings (Cluster 2's Rating 3.76 and Cluster 3's 3.68, with declining recent satisfaction), suggesting that branches in the fast-food sector face challenges in achieving top-tier customer satisfaction, especially when dealing with large customer volumes. In contrast, the high-performing Cluster 6 has a much more balanced sector distribution, with significant representation of coffeehouses (23%) and patisserie/bakery outlets (26%) alongside fast-casual restaurants (41%), and only a small fraction of fast-food (10%). Coffeehouse and bakery/patisserie branches generally emphasize product quality and a relaxed customer experience, which might contribute to the higher average ratings observed in Cluster 6. Meanwhile, Cluster 5, which showed improving high performance, is composed predominantly of fast-casual restaurants (75% of its branches) with a minor presence of other types; fast-casual establishments typically offer a blend of quick service and higher quality, which may explain their ability to garner strong reviews in that cluster. On the other hand, Cluster 4 and Cluster 7, both low-performing, are centered on similar sectors as some high performers but with different outcomes. Cluster 4 has a mix of fast-casual (46%) and fast-food (39%)

branches, and Cluster 7, despite being 45% fast-casual and having a notable share of patisseries (24%) and fast-food (23%), still manages to underperform drastically. The poor results of Clusters 4 and 7, even though they include many fast-casual restaurants (a category that can succeed as seen in Cluster 5), hint that the issues are likely brand-specific or operational in nature. It may be that certain chains or branches in those clusters are not executing the fast-casual concept effectively, resulting in customer dissatisfaction. In summary, sector composition provides context to cluster performance: fast-food-heavy clusters tend to have large volumes but struggle with ratings, whereas clusters with more coffeehouse and bakery elements (and well-run fast-casual concepts) tend to excel. Nonetheless, high or low performance is not solely determined by sector—management quality and customer service within each sector play a critical role, as evidenced by the stark contrast between Cluster 5 and Cluster 7 despite both being dominated by fast-casual outlets.

5.5 Conclusion

This study proposed a data-driven clustering approach to evaluate the service performance of brand chain branches in Türkiye's rapidly expanding foodservice industry. Utilizing customer review data sourced from Google Reviews, the analysis focused on both cumulative and recent performance indicators, including review frequency and average ratings. By applying the k-means clustering algorithm, branches were segmented into distinct performance groups, offering nuanced insights into operational outcomes. These clusters were then examined in relation to contextual characteristics such as sector, location, and urbanization level.

Bringing together the performance outcomes and branch characteristics, we observe some clear contrasts between the high- and low-performing clusters that offer valuable insights for both researchers and practitioners. High-performing clusters (5 and 6) are defined by excellent customer satisfaction and stable or improving recent performance, and they illustrate that diverse branch profiles can succeed—these clusters include a mix of branch types and are not confined to a single sector or location type. For instance, Cluster 6's branches span coffeehouses, restaurants, and bakeries in both mall and non-mall settings, yet consistently deliver high ratings (as shown in Table 5.3). This suggests that strong operational practices and service quality standards can be implemented across different formats and locations to achieve uniformly positive outcomes. In contrast, the low-performing clusters (particularly Clusters 4 and 7) highlight contexts where service strategies may be failing. These clusters concentrate a large number of mall-based, urban branches that should have access to high customer traffic, yet their very low recent ratings (around 1.87) indicate systemic issues such as poor service experience, inadequate staffing or training, or mismatch between customer expectations and delivery. The fact that Cluster 7's branches are predominantly fast-casual restaurants in prime locations but still underperform significantly underscores that simply having a good location or a popular service format is not enough, consistent execution and customer satisfaction management are critical.

From an academic perspective, these results demonstrate the utility of a data-driven clustering approach in uncovering heterogeneity in service performance. The distinct clusters reveal how combinations of review volume and rating can segment branches into meaningful groups that correlate with tangible business characteristics. This contributes to the literature by empirically validating that online review metrics can serve as effective indicators to classify service units (branches) and that such classifications align with real-world operational contexts (like location and sector). For example, the identification of a cluster of high-traffic but declining-satisfaction branches (Cluster 3) provides a case for further study into capacity management and service consistency in extremely busy outlets. Likewise, the presence of a large cluster of consistently high-performing branches (Cluster 6) invites investigation into what managerial practices or customer service principles are common in those units, contributing to their success across varied conditions.

From a practical standpoint, the clustering outcomes offer actionable insights for operational strategy and service quality improvement. Managers of foodservice chains can use these findings to benchmark their branches and prioritize interventions. Specifically, branches falling into the low-performing clusters (e.g., the branches in Cluster 7) should be targeted for immediate improvement efforts, this may involve retraining staff, overhauling service processes, or enhancing the customer experience to address the root causes of dissatisfaction. The analysis suggests that many of these underperforming branches are in shopping malls, so it may be beneficial to review how those branches operate within the mall environment (for instance, their peak-hour staffing, engagement with mall promotions, or competition with neighboring food outlets). On the other hand, the high-performing clusters provide best-practice models. Cluster 6, with its widespread success, likely encompasses effective managerial strategies and customer-centric practices that could be replicated in weaker branches. Similarly, the improvements seen in Cluster 5 imply that changes implemented in those branches (such as menu updates, loyalty programs, or service training that occurred recently) have had a positive impact, these initiatives could be studied and extended to other branches in the network. Finally, the moderate clusters (1 and 2) indicate branches that are doing reasonably well but have potential to improve; proactive measures here (for example, addressing the slight downward trend in Cluster 2's recent ratings by investigating common complaints) could prevent these mid-tier branches from sliding into poor performance. In essence, the clustering results enable a chain to adopt a segment-specific strategy: bolstering the strong clusters, learning from them, and lifting the weaker ones by tailoring improvement programs to the specific issues identified in each cluster profile.

In conclusion, the results of the cluster analysis paint a comprehensive picture of branch performance diversity within the foodservice chains. By quantitatively grouping branches and then examining their qualitative attributes, we gain a dual perspective: descriptively, we see what performance levels exist and which clusters stand out at the extremes; analytically, we can associate those performance levels with underlying factors such as location type and business model. This integrated understanding is crucial for academics studying service management, as it provides real-world evidence of performance drivers, and for practitioners seeking to enhance service quality and customer satisfaction across a large network of service outlets. The clear differences highlighted between high- and low-performing clusters (as shown in Tables 5.3 and 5.4) serve as a roadmap for targeted improvements and further research into the determinants of service excellence in the foodservice industry.

While this study provides a valuable and interpretable framework for assessing service performance using a data-driven clustering approach, there are several limitations that also suggest avenues for future research. The clustering was based on four core performance indicators derived from Google Reviews, which, while robust and interpretable, may not capture all dimensions of service quality. Future research could explore the inclusion of additional features, such as temporal trends, textual sentiment from reviews, or operational data, and experiment with alternative clustering algorithms (e.g., DBSCAN, hierarchical clustering, or Gaussian Mixture Models) to validate or complement the current findings. Furthermore, incorporating automated cluster evaluation metrics, such as the silhouette score or Davies-Bouldin index, would strengthen the methodological rigor and support more scalable applications of this approach in different contexts. Importantly, to ensure the ongoing relevance and actionability of the insights, this type of analysis should be repeated periodically by relevant companies, enabling them to monitor performance trends over time and adjust their strategies accordingly.

Appendix

See Table 5.5.

Table 5.5 Summary of related literature

Article	Aim	Methods	Data
[32]	Analyse environment-related hotel reviews using text mining to enhance hotel services and customer satisfaction	Data scraping, machine learning classification, and sentiment analysis on reviews from otelpuan.com	211,790 reviews from 103 hotels in Antalya, classified and analysed for environmental content and sentiment
[24]	Conduct cross-sector meta-analysis of online reviews to improve service quality using quality management tools	Meta-analysis using quality management tools (Juran's trilogy, CTQ, ERVQUAL), bigram / trigram analysis, and E-LDA	32,400,048 reviews across healthcare, hotels, airlines, restaurants, online games, and online food/grocery delivery
[23]	Assess hospitality service quality using machine learning and NLP on customer reviews to identify key aspects	Natural Language Processing (NLP), Latent Dirichlet Allocation (LDA), FastText, and dimensionality reduction techniques	Customer reviews from Booking.com for hotels in Bogotá and Madrid, totalling 108,563 reviews for Bogotá and 498,361 for Madrid
[11]	Understand customer preferences and satisfaction in small fast-food businesses compared to national chains using Google Reviews data	Text analysis, sentiment analysis, and multilevel multinomial models on 82,598 Google reviews	82,598 Google reviews from 235 chain and 387 small business restaurants in Upstate New York cities: Albany, Buffalo, Rochester, and Syracuse
[15]	Evaluate airport service quality using Google reviews, cross-validate ASQ survey results, and compare Google reviews with ASQ attributes	Employ sentiment analysis, topic modelling, and pairwise comparisons between Google reviews, Google star ratings, and ASQ ratings	Analyse 42,137 Google reviews, compare with ASQ survey data, and validate using sentiment scores, star ratings, and extracted topics
[27]	Utilize machine learning on Google Maps Reviews to enhance participatory planning for public parks in Istanbul	Data scraping, XGBoost classification, Shapley value analysis, and keyword extraction from Google Maps Reviews	Google Maps Reviews data for four parks in Istanbul, focusing on reviews written in Turkish, with a total of 4,437 reviews analysed
[21]	Measure airport service quality using machine learning algorithms and sentiment analysis on passenger reviews	Data scraping, text preprocessing (TF-IDF), sentiment analysis using Logistic Regression, XGBoost, SVM, Random Forest, and Naïve Bayes	38,105 reviews from SKYTRAX, covering 600 airports, including ratings for queuing times, cleanliness, seating, signs, food, shopping, Wi-Fi, and staff
[33]	Investigate changes in airport service quality (ASQ) during COVID-19 using crowdsourced sentiment analysis of Google Maps reviews	Data collection, sentiment analysis, and topic ontology development; correlation and regression analysis on reviews of 98 busiest U.S. airports	642,546 Google Maps reviews from 98 busiest U.S. airports, split into pre-COVID-19 (138,411 reviews) and post-COVID-19 (462,135 reviews)

(continued)

Table 5.5 (continued)

Article	Aim	Methods	Data
[26]	Evaluate customer reviews of The Clink prison training restaurants to understand dining experiences and impact on perceptions	Interpretive case study using Leximancer 5.0 to analyse 3,951 TripAdvisor reviews for thematic content	3,951 TripAdvisor reviews of The Clink restaurants in four UK prisons, analysed for themes like food, visit, and prison
[5]	Examine effects of restaurant attributes on customer experience using Google Maps reviews of various restaurant types	Quantitative analysis, sentiment analysis with VADER algorithm, logistic regression, and lexical salience-valence analysis on 935,386 reviews	935,386 Google Maps reviews from 5,010 restaurants in London, Birmingham, and Manchester, analysing food, service, atmosphere, and value attributes
[20]	Compare performance of SVM and RF algorithms for sentiment analysis on tourism reviews to derive strategic decisions	Data collection, preprocessing, feature extraction (TF-IDF, N-Gram, Word2Vec, POS), SVM and RF models, evaluated with 10-Fold Cross Validation	11,000 reviews (5,500 positive, 5,500 negative) from TripAdvisor, HolidayIQ, Twitter, Facebook, and Google Reviews for tourist places in Chhattisgarh and Maharashtra, India
[29]	Develop an integrated framework to analyse and visualize Hilton hotel ratings and reviews from TripAdvisor using sentiment analysis	Data scraping, sentiment-sensitive tree construction, convolution tree kernel classification, aspect extraction, visual analytics using Google Trends data	634,277 reviews from 749 Hilton hotels in the U.S., covering ratings for Value, Location, Sleep Quality, Rooms, Cleanliness, Service, Check-in/Front Desk, and Business Service
[19]	Diagnose service quality using customer reviews through sentiment and gap analyses for better quality management	Sentiment analysis, customer review-based gap analysis, service-feature hierarchy construction, sentiment word dictionary creation	579 customer reviews from iPhone Apps Plus for a mobile navigation service, analysed for perceptions and expectations at the service-feature level
[14]	Use topic modelling to measure airline service quality from unstructured online review data for better customer satisfaction insights	Structural Topic Models (STM) combining textual analysis of reviews with covariates like rating scores, sentiment analysis, and Bayesian inference	557,208 TripAdvisor reviews, including information on flight details, service aspects (e.g., seat comfort, customer service), and passenger metadata
[16]	Compare restaurant inspection report results with star ratings on Yelp and Google Reviews to assess public perceptions of food safety	Secondary data collection from Fraser Health inspection reports and Yelp/Google Reviews; chi-square tests and correlation analyses	170 restaurant inspection reports from Fraser Health, Yelp, and Google Reviews star ratings for restaurants in Surrey, Burnaby, and Abbotsford, BC

(continued)

Table 5.5 (continued)

Article	Aim	Methods	Data
[17]	Analyse online- generated reviews (OGRs) to complement official restaurant inspection procedures and improve food safety	Data scraping with Python, sentiment analysis, correlation analysis between OGRs from TripAdvisor and health inspection scores	50,618 TripAdvisor reviews and 1,045 restaurant health inspection results from Manhattan, New York, analysed for correlation
[12]	Investigate the relative importance of online review attributes in consumer restaurant selection, focusing on food quality, service quality, atmosphere, and price	Choice-based conjoint (CBC) analysis, quantitative approach, MTurk platform for data collection, and statistical analysis of review attributes	353 responses from Amazon's Mechanical Turk (MTurk), analysing online reviews for attributes like food quality, service quality, atmosphere, price, and overall restaurant rating
[13]	Examine how reviewer trustworthiness impacts business patronage, developing a model to predict reviewer trustworthiness	Data scraping, regression analysis, logistic regression for classification, using Yelp data and source credibility theory	Yelp data including 2.2 million reviews, 77,000 businesses, and 525,000 users, with metrics for reviews, ratings, and reviewer characteristics
[8]	Identify the impact of online consumer reviews and management responses on hotel performance using a panel data analysis	Panel data analysis on 843 hotels, examining review ratings, volume, variation, and management responses; econometric modelling	843 hotels' reviews from TripAdvisor, matched with quarterly hotel performance data from a local revenue comptroller, covering 10 quarters (2009–2011)
[9]	Analyse customer satisfaction and dissatisfaction with hotel products and services using online reviews to identify key attributes	Text mining, Latent Semantic Analysis (LSA), and regression analysis to examine customer reviews from Booking.com	3,596 online reviews from Booking.com, covering various hotels in the 100 largest U.S. cities, analysed for customer perceptions and satisfaction attributes
[10]	Study the impact of marketing decisions on eWOM and business performance for Spanish rural lodging establishments	Data collection from Toprural, surveys of lodging owners, regression analysis on eWOM and performance variables	202 Spanish rural lodging establishments, Toprural reviews and ratings, and survey data on business performance
[27]	Develop machine learning models to predict the helpfulness of consumer reviews using textual features like polarity and reading ease	Data scraping, feature extraction (polarity, subjectivity, entropy, reading ease), machine learning models for prediction, and evaluation	622,494 Amazon reviews across books, baby products, and electronics, including fields like review text, helpfulness votes, and review metadata

References

1. Mordor Intelligence: Turkey foodservice market sıze & share analysıs -growth trends & forecasts up to 2029 (2024)
2. Mamalis, S.: Critical success factors of the food service industry. J. Int. Food Agribus. Mark. **21**, 191–206 (2009). https://doi.org/10.1080/08974430802589709
3. Kandampully, J., Duddy, R.: Competitive advantage through anticipation, innovation and relationships. Manag. Decis. **37**, 51–56 (1999). https://doi.org/10.1108/00251749910252021
4. Patil, D.R., Rane, N.L.: Customer experience and satisfaction: importance of customer reviews and customer value on buying preference. Int. Res. J. Mod. Eng. Technol. Sci. **5**, 3437–3447 (2023)
5. Mathayomchan, B., Taecharungroj, V.: "How was your meal?" Examining customer experience using Google maps reviews. Int. J. Hosp. Manag. **90**, 102641 (2020). https://doi.org/10.1016/j.ijhm.2020.102641
6. Kim, B., Velthuis, O.: From reactivity to reputation management: online consumer review systems in the restaurant industry. J. Cult. Econ. **14**, 675–693 (2021). https://doi.org/10.1080/17530350.2021.1895280
7. Ching, M.R.D., de Dios Bulos, R.: Improving restaurants' business performance using Yelp data sets through sentiment analysis. In: Proceedings of the 3rd International Conference on E-commerce, E-Business and E-Government, pp. 62–67 (2019)
8. Xie, K.L., Zhang, Z., Zhang, Z.: The business value of online consumer reviews and management response to hotel performance. Int. J. Hosp. Manag. **43**, 1–12 (2014). https://doi.org/10.1016/j.ijhm.2014.07.007
9. Xu, X., Wang, X., Li, Y., Haghighi, M.: Business intelligence in online customer textual reviews: understanding consumer perceptions and influential factors. Int. J. Inf. Manag. **37**, 673–683 (2017). https://doi.org/10.1016/j.ijinfomgt.2017.06.004
10. Nieto, J., Hernández-Maestro, R.M., Muñoz-Gallego, P.A.: Marketing decisions, customer reviews, and business performance: the use of the Toprural website by Spanish rural lodging establishments. Tour. Manag. **45**, 115–123 (2014). https://doi.org/10.1016/j.tourman.2014.03.009
11. Yalcinkaya, B.: Customer preferences in small fast-food businesses: a multilevel approach to google reviews data (2020)
12. Gunden, N.: How online reviews influence consumer restaurant selection (2017). https://www.proquest.com/dissertations-theses/how-online-reviews-influence-consumer-restaurant/docview/1891353397/se-2?accountid=25089
13. Banerjee, S., Bhattacharyya, S., Bose, I.: Whose online reviews to trust? Understanding reviewer trustworthiness and its impact on business. Decis. Support Syst. **96**, 17–26 (2017). https://doi.org/10.1016/j.dss.2017.01.006
14. Korfiatis, N., Stamolampros, P., Kourouthanassis, P., Sagiadinos, V.: Measuring service quality from unstructured data: A topic modeling application on airline passengers' online reviews. Expert Syst. Appl. **116**, 472–486 (2019). https://doi.org/10.1016/j.eswa.2018.09.037
15. Lee, K., Yu, C.: Assessment of airport service quality: a complementary approach to measure perceived service quality based on google reviews. J. Air Transp. Manag. **71**, 28–44 (2018). https://doi.org/10.1016/j.jairtraman.2018.05.004
16. Kong, E., Heacock, H.: Comparison of restaurant inspection report results and its corresponding star ratings on Yelp and Google Reviews. BCIT Environ. Public Heal. J. (2020)
17. Grimaldi, D., Collins, C., Acosta, S.G.: Dynamic restaurants quality mapping using online user reviews (2021)
18. Yusupova, N.I., Bogdanova, D.R., Komendantova, N.P.: Artificial intelligence tools for analyzing emotionally colored information from customer reviews in the service sector. IOP Conf. Ser. Mater. Sci. Eng. **1069**, 12013 (2021). https://doi.org/10.1088/1757-899X/1069/1/012013

19. Song, B., Lee, C., Yoon, B., Park, Y.: Diagnosing service quality using customer reviews: an index approach based on sentiment and gap analyses. Serv. Bus. **10**, 775–798 (2016). https://doi.org/10.1007/s11628-015-0290-1
20. Sahu, M.K., Selot, S.: Support vector machine and random forest machine learning algorithms for sentiment analysis on tourism reviews: a performance analysis. I-manager's J. Comput. Sci. **9** (2021)
21. Homaid, M.S., Moulitsas, I.: Measuring airport service quality using machine learning algorithms. In: Proceedings of the 6th International Conference on Advances in Artificial Intelligence, pp. 8–14 (2022)
22. Barik, K., Misra, S.: Analysis of customer reviews with an improved VADER lexicon classifier. J. Big Data. **11**, 10 (2024). https://doi.org/10.1186/s40537-023-00861-x
23. Vargas-Calderón, V., Moros Ochoa, A., Castro Nieto, G.Y., Camargo, J.E.: Machine learning for assessing quality of service in the hospitality sector based on customer reviews. Inf. Technol. Tour. **23**, 351–379 (2021). https://doi.org/10.1007/s40558-021-00207-4
24. Rajendran, S., Baskar, S.: Improving quality of service industries using online customer reviews: cross-sector study. In: 2021 International Conference on Electrical, Computer, Communications and Mechatronics Engineering (ICECCME), pp. 1–6 (2021)
25. Shin, J., Joung, J., Lim, C.: Determining directions of service quality management using online review mining with interpretable machine learning. Int. J. Hosp. Manag. **118**, 103684 (2024). https://doi.org/10.1016/j.ijhm.2023.103684
26. Gebbels, M., McIntosh, A., Harkison, T.: Fine-dining in prisons: online TripAdvisor reviews of the clink training restaurants. Int. J. Hosp. Manag. **95**, 102937 (2021). https://doi.org/10.1016/j.ijhm.2021.102937
27. Singh, J.P., Irani, S., Rana, N.P., Dwivedi, Y.K., Saumya, S., Kumar Roy, P.: Predicting the "helpfulness" of online consumer reviews. J. Bus. Res. **70**, 346–355 (2017). https://doi.org/10.1016/j.jbusres.2016.08.008
28. Ma, B., Wong, Y.D., Teo, C.-C., Wang, Z.: Enhance understandings of online food delivery's service quality with online reviews. J. Retail. Consum. Serv. **76**, 103588 (2024). https://doi.org/10.1016/j.jretconser.2023.103588
29. Chang, Y.-C., Ku, C.-H., Chen, C.-H.: Social media analytics: extracting and visualizing Hilton hotel ratings and reviews from TripAdvisor. Int. J. Inf. Manag. **48**, 263–279 (2019). https://doi.org/10.1016/j.ijinfomgt.2017.11.001
30. Akkaya, M., Özlem, Ö., Tepe, E.: A machine learning application to Google Maps Reviews as a participatory planning tool. Int. J. Urban Sci. **28**, 379–402 (2024). https://doi.org/10.1080/12265934.2024.2320916
31. Rykov, A., Amorim, R.C.D., Makarenkov, V., Mirkin, B.: Inertia-based indices to determine the number of clusters in K-means: an experimental evaluation. IEEE Access **12**, 11761–11773 (2024). https://doi.org/10.1109/ACCESS.2024.3350791
32. Gürbüz, M., Sürmeli, D., Taşkın, K., Cebeci, H.İ: Otellere için paylaşılan çevre ile alakalı yorumların metin madenciliği ile analizi: antalya otelleri üzerine bir araştırma. Bus. Manag. Stud. An Int. J. **12**, 218–239 (2024)
33. Li, L., Mao, Y., Wang, Y., Ma, Z.: How has airport service quality changed in the context of COVID-19: a data-driven crowdsourcing approach based on sentiment analysis. J. Air Transp. Manag. **105**, 102298 (2022). https://doi.org/10.1016/j.jairtraman.2022.102298

Chapter 6
Integrating Quality Management 4.0 with AI and Machine Learning

Leilani Uttenreither, Mahesh Raisinghani, and Fen Wang

Abstract This manuscript reviews recent literature on project quality management, Quality 4.0, and strategic planning, and synthesizes its findings to offer a novel framework of integrating Quality 4.0 principles to strategic planning processes. While strategic planning remains critical for shaping organizational objectives, there is limited research on leveraging quality management practices in this context. The methodology involves a qualitative literature review, outlining quality management evolution and discussing Quality 4.0's use of AI and machine learning. Gaps are identified regarding integrating quality with strategic planning processes. A novel framework is proposed for "Strategic Planning 4.0" that incorporates Quality 4.0 principles of automation, data analytics and machine learning into the strategic planning process. Implications for enhancing strategic planning through Quality 4.0, limitations, and areas for future research are discussed, positioning organizations for success in the Industry 4.0 era.

Keywords Data science · Analytics · AI/Machine Learning (AI/ML) · AI/ML projects · Quality 4.0

L. Uttenreither
University of Maryland Global Campus, College Park, MD, USA
e-mail: leilanipearl@gmail.com

M. Raisinghani
Texas Woman´S University, Denton, TX, USA
e-mail: mraisinghani@twu.edu

F. Wang (✉)
Central Washington University, Ellensburg, WA, USA
e-mail: fen.wang@cwu.edu

M. Mora et al. (eds.), *Engineering and Management of Data Science, Analytics, and AI/ML Projects*, Intelligent Systems Reference Library 282,
https://doi.org/10.1007/978-3-032-06889-7_6

6.1 Introduction

Strategic planning is an important and prevailing process for shaping organizational strategic objectives [1] as it aligns vision, shares responsibility, and clarifies desired outcomes [2]. Despite the critical role of strategic planning, there remains a lack of research on how quality management practices can effectively be integrated into this important process. Recently, there has been a growth in quality management literature that focuses on "Quality 4.0," which leverages new technologies such as artificial intelligence and machine learning with quality assurance [3]. While there is a general consensus on best practices in producing a formal strategic plan [2, 4, 5], there is a gap in research on how to integrate modern quality management tools (i.e., Quality 4.0) in the process of strategic planning. Yet, in today's rapidly evolving business landscape, managers seek solutions to enhance strategic planning with data-driven decisions and continuous adaptation [1].

Strategic planning is an important management tool, however, if the process is done poorly, the plan can be ineffective and burdensome [6]. Recent literature on quality management suggests the importance of integrating automation and machine learning with quality assurance to remain competitive in the fourth industrial age, also known as Industry 4.0, that is defined by these technological advancements [7]. However, Quality 4.0 is an emerging field of knowledge. This manuscript examines recent literature that focuses on Quality 4.0 in the context of project quality management and strategic planning. Then it synthesizes the findings to provide a novel framework for leveraging these new technologies and propelling the evolution toward "Strategic Planning 4.0." Executives, managers, strategic planners, and quality management professionals actively engaged in the strategic planning process will find this research compelling, as it provides insights into integrating these technologies, fostering innovation, and achieving success in the dynamic era of Industry 4.0 [7].

This manuscript represents a significant and valuable contribution to project management research. Notably, it pioneers the application of quality management methods and "Quality 4.0" principles to the strategic planning process. By doing so, it provides essential insights for organizational leaders who aim to leverage emerging technology and streamline their strategic planning processes. As organizations navigate the complexities of Industry 4.0, the incorporation of Quality 4.0 principles becomes critical across all facets of organizational strategy, including strategic planning processes.

6.2 Methodology

The methodology of this manuscript consists of a qualitative literature review of research in the domains of project management, quality management, and strategic planning within the past four years. Additional background for the purposes of describing relevant theoretical frameworks and assumptions relies on some older literature. This qualitative literature review aims to explore and synthesize existing scholarly works, focusing on qualitative data and interpretations to gain a comprehensive understanding of the applications of quality management in strategic planning. The search strategy was mostly conducted through the online library offered by the University of Maryland Global Campus with the key phrases "strategic plan", "quality management", "TQM", and "Quality 4.0" and limited to the years 2020–2024. Some research was supplemented through search engines and prompt engineering with the same keywords. The process entailed many rounds of searching and screening to ensure comprehensive coverage of research literature. Each identified source was screened through a review of the manuscript's research question and its relevance to the topic. Recurring themes and concepts were extracted from the selected research to generate new insights and understandings.

6.3 Literature Review

Quality management is frequently overlooked and undervalued, however, it plays a critical role in organizational excellence [8]. The literature review in this manuscript provides the historical context of quality management theory and strategic planning, and discusses emerging research on "Quality 4.0" from the past four years. Soltani [9] highlights that in the current highly competitive global market, companies must prioritize both quality and innovation to ensure their survival. Amid the dynamic business landscape of the 2020s, business leaders must grasp the importance of a proactive, high-quality approach when formulating organizational strategies.

6.3.1 Overview of Quality Management

The PMBOK [10] defines quality management as breaking down into three domains: planning, managing, and controlling. It is worth noting that developing a quality management plan is crucial because a structured approach ensures that high-quality deliverables are consistently produced throughout the project life cycle [10]. The quality management plan defines the quality metrics and quality control activities so that the project deliverables meet the project requirements [10].

Lotich [11] discussed the Total Quality Management (TQM) system and emphasized that organizations should clearly communicate their vision, mission, and values to stakeholders to foster alignment with strategic objectives. Customized Critical Success Factors (CSFs) are crucial for tracking progress, and relevant measurements and metrics should be established [11]. Additionally, creating a customer-focused culture that prioritizes satisfaction and involving all employees in the quality management system contributes to successful TQM implementation [11].

Table 6.1 Quality philosophy

(a) Statistical quality control (SQC)	(b) Total Quality Management (TQM)	(c) Six sigma	(d) Design for Six Sigma (DFSS)	(e) Quality 4.0
Controlling Specification Production Inspection	Managing Plan Do Check Act	Reactive Define Measure Analyze Improve Control	Proactive Define Measure Analyze Design Optimize Verify	Predictive Identify Accessorize Discover Learn Predict Redesign Relearn

"Evolution of quality, problem-solving strategy by quality philosophy." From Escobar (2021)

Escobar [12] offered a comprehensive history of changes in quality management philosophy. It began in manufacturing industries with the use of statistical tools (Pareto charts, histograms, etc.) to measure and control quality. This later evolved to TQM (plan, do, check, act) and Six Sigma (define, measure, analyze, improve, control). However, Escobar [12] noted that recent advancements in technology now require a predictive element in quality management, which is based on quality curve theory, or Quality 4.0. While the definitive activities of Quality 4.0 are still in the theoretical stages of research, Escobar [12] suggests a predictive framework of seven steps: identify, ascensorize (defined as deploying sensors and observe), discover, learn, predict, redesign, and relearn (Table 6.1).

6.3.2 *Quality 4.0 with AI and ML*

Artificial Intelligence (AI) and Machine Learning (ML) are fields that aim to enable machines to perform tasks that typically require human intelligence. ML, a subset of AI, focuses on algorithms that allow systems to learn from data without explicit programming. More specifically, ML is broadly categorized into supervised learning, unsupervised learning, and reinforcement learning, each differing in how models learn from input data. The theoretical background of AI/ML encompasses various areas, including statistical learning theory, computational learning theory, and the study of fundamental principles underlying AI systems. Accordingly, AI/ML technologies serve as foundational tools for data-driven automation and intelligent decision-making in modern systems. In the 21st century, AI is criticized as all-powerful, biased and controlled by big-tech oligarchs [13]. Inevitably, artificial intelligence and machine learning are profoundly influencing various facets of project management—an impact that is becoming increasingly apparent in the era of Industry 4.0.

According to research by Alzahrani, Bahaitham, Andejany, and Elshennawy [7], a fourth industrial revolution is transforming the world, also referred to as "Industry

4.0”. The original Industrial Revolution (1.0) began in the late 1700s with the discovery of coal and steam power. Industry 2.0 began in the late 1800s with the advancements of oil, electricity, and mass production. In the late 1900s, Industry 3.0 was sparked by nuclear energy and electronics. Finally, Industry 4.0 is defined by advancements in digital technologies, machine learning, and artificial intelligence. These developments have led to the emergence of quality curve theory, or Quality 4.0, as quality management has changed as reflected by technological advancements [7]. Organizations must evaluate their readiness for Quality 4.0 adoption, considering enabling technologies, big data capabilities, skilled workers, collaboration, and leadership support [7].

Liu et al. [14] discussed Quality Curve Theory and suggested that Quality 4.0 activities fall into three categories: design, manufacturing, and service. It is important to note that the main premise of this theory is that the process design and service end contribute significantly more to quality than manufacturing (or production) quality in and of itself. In other words, organizations need to emphasize design and service activities, or the beginning and the end of the process, to thrive in the current climate [14].

Liu et al. [14] also noted that the difference between TQM and Quality 4.0 is that while TQM focuses on the alignment of activities with customer requirements and expectations, Quality 4.0 emphasizes a predictive and proactive approach through automation, advanced data analysis, and result integration. In other words, TQM looks to improve upon past performance, while Quality 4.0 aims to reduce defects through technology and automation.

Maganga and Taifa [15] describe Quality 4.0 as a modern quality management approach that leverages Industry 4.0 technologies, integration, and digitalization. They noted that publications on Quality 4.0 began in 2016 and surged in 2020 and 2021, with India leading in research output and little output from the United States. Maganga and Taifa [15] suggest that the framework of Quality 4.0 is based on four pillars: TQM principles, Industry 4.0 technologies, big data management, and real-time quality management. Finally, their research suggested that the top three motivating factors for adopting Quality 4.0 is for organizations to provide the required big data tools, assistance in decision-making, and incentives for continuous improvement.

Emblemsvåg [16] emphasized the importance of replacing manual data collection with automated methods such as QR codes, barcodes, and RFID chips, which can enhance data quality, security, and centralization through cyber security and cloud computing technologies. As noted earlier, Escobar et al. [12] also stated that Quality 4.0 integrates Industry 4.0 technologies, defines digitalization, and emphasizes enablers like big data capabilities and leadership support. Furthermore, Emblemsvåg [16] and Escobar et al. [12] agreed on the unreliability of manual or visual inspections, which can be subject to bias, and the need to replace manual methods with automation to the extent possible.

A systematic literature review conducted by De Oliveira, Alvelos, and Rosa [3] revealed a significant global increase in scholarly publications related to Quality 4.0 since 2020, underscoring its paramount importance in global quality management.

De Oliveira et al. [3] illustrated the evolution of quality management to align with Industry 4.0, highlighting the integrated approach involving new technologies and methodologies. De Oliveira et al. [3] emphasized key aspects include motivations (customer satisfaction, productivity, cost and time savings), barriers (initial costs, employee skills, lack of resources), and readiness factors (leadership, culture, data analytics, training, and scalability). Quality 4.0 professionals must possess a diverse skill set, including proficiency in technology utilization, big data analysis, effective communication, leadership, creative problem-solving, adaptability to change, knowledge of IT and production processes, and a commitment to continuous lifelong learning [3]. These skills are essential for achieving success within a Quality 4.0 system, which combines automation and real-time data analysis while integrating traditional tools and methodologies [3].

6.3.3 Overview of Strategic Planning

Strategic planning continues to be the prevailing method for shaping strategies [1]. Cote [2] defined strategic planning as an ongoing organizational process that documents a business's goals and objectives by prioritizing efforts, allocating resources, aligning stakeholders with organizational goals, and ensuring data-backed reasoning. While a strategic plan is time-bound with a beginning and an end, the strategic planning process should be characterized by continuous learning and improvement. The benefits of strategic planning include a unified vision, enhanced responsibility and accountability, and improved business outcomes [2].

Edwards et al. [4] offered a case study that exemplifies the traditional components of a successful strategic plan implementation. The study emphasized the strategic importance of strengthening partnership networks and the initial challenges faced in executing their strategy. Edwards et al. [4] noted that strategy is a process, not an event, and that it is imperative to develop a common vision, "SMART" goals, and a cohesive decision-making process. Edwards et al. [4] also recommended standardization, prioritization, collaboration, accountability, and the use of dashboards for initiatives, implementation status, and impact assessment.

George [1] noted that effective implementation of strategic plans is shaped by the interplay of three key factors: people, process, and plan (often referred to as the 3Ps). George [1] advised assembling a diverse team with creative thinking to drive strategic initiatives. Stakeholder involvement at all levels ensures alignment and commitment. Data-driven decisions, continuous adaptation, risk mitigation, and high-quality plans are essential. Strategic planning is not one-size-fits-all. Variation in people, process, and plan impacts implementation success.

In a compelling case study, El Zein Badawi [5] explored the successful integration of quality assurance and strategic planning within a Sudanese higher education institution (HEI). The study emphasized that these two critical aspects should not exist in isolation. By strategically aligning quality assurance efforts with the university's vision, mission, and objectives, the HEI achieved notable improvements in organizational performance and student outcomes. El Zein Badawi [5] underscored the importance of clear, validated, benchmarked, and transparent key performance indicators (KPIs). Additionally, the study highlighted that while quality assurance provides insights into past performance, integrating it with strategic planning is essential for HEIs to thrive in a dynamic enterprise environment.

6.3.4 Gaps in Literature

Existing research on the integration of quality management and strategic planning reveals several gaps that warrant further exploration. First, existing literature on strategic planning tends to emphasize development and implementation, often overlooking an assessment of the quality of the strategic planning process. Second, while the PMBOK [3] offers valuable guidance on project quality management, empirical research on Quality 4.0 is an emerging field with limited established guidance or frameworks for successful implementation within organizations [4]. Third, although the potential advantages of incorporating Quality 4.0 into organizational management are recognized, empirical studies on its concrete impact or organizational performance remain limited, likely due to its status as an emerging research field. Lastly, despite a growing body of conceptual literature on Quality 4.0, empirical research examining its practical application and impact on strategic planning processes within organizations remains elusive. This manuscript proposes a framework for organizations to enhance strategic planning by integrating Quality 4.0 principles, ensuring alignment with the dynamic Industry 4.0 context.

6.4 Proposed Process for Strategic Planning 4.0

The PMBOK [17] highlights that quality activities aim to align delivered outcomes with customer and stakeholder objectives, minimizing resource waste, and increasing the likelihood of achieving desired results. The complexities arise from human behavior, system interdependencies, uncertainty, and technological innovation. Yet, Soltani [9] noted that quality management is foundational for product and process innovation, and that organizations must prioritize best practices in quality management to drive innovation and performance.

Quality 4.0 activities, as highlighted by Krivokuca [8], emphasize digital technologies, data-driven decision-making, smart manufacturing, risk management, collaboration, and continuous improvement to enhance quality. While strategic planning remains the dominant approach for shaping strategies [1], effective implementation hinges on the interplay of three key factors: people, process, and plan (often referred to as the 3Ps). Aligning "Strategic Planning 4.0" with Quality 4.0 is critical for organizational success in the Industry 4.0 era, yet a consensus on a practical framework remains elusive.

Escobar [12] proposed a practical framework for Quality 4.0, termed "the evolution of problem-solving," which is currently in its early development stage. The objective is to enhance process quality by incorporating predictive analytics and automation, while minimizing reliance on manual updates and inspections. The below process for Strategic Planning 4.0 is modeled on "the evolution of problem-solving" [12] and organized by the Quality 4.0 principles of people-process-technology [18] and the people-process-plan approach by [10]. Figure 6.1 presents a visual framework illustrating the Proposed Process for Strategic Planning 4.0, which outlines seven iterative stages—Identify, Accessorize, Discover, Learn, Predict, Redesign, and Relearn—each integrating people, processes, technology, and planning activities to drive continuous improvement through data-driven decision-making and machine learning (Table 6.2).

Proposed Process for Strategic Planning 4.0 modeled on "The Evolution of Problem Solving" Escobar [12] by Leilani Uttenreither.

6.4.1 Potential Application Guidelines

While traditional quality management focuses on continuous improvement through production feedback, Quality 4.0 prioritizes proactive design process through automation and machine learning [14]. Escobar [12] offered a problem-solving strategy that could provide a possible framework to apply automation to the strategic planning process. However, Escobar [12] also noted challenges in adopting the tools including a lack of understanding and trust, a lack of applicability for all situations, machine learning is not appropriate for root-cause analysis, and practical complications with the final relearning component. Krivokuca [8] cautioned that the involvement of people remains critical to processes, and it is imperative to emphasize that artificial intelligence will not replace people. Overall, while challenges persist in integrating big data and automation into strategic planning, the benefits of minimizing manual updates and leaning into automation are evident.

Fernández-Loría et al. [19] examine counterfactual explanations for explaining the decisions made by model-based AI systems. They describe three detailed case studies that compare the counterfactual approach with Shapley additive explanations to illustrate conditions under which counterfactual explanations explain data-driven decisions better than importance weights. This counterfactual approach could be helpful in the context of integrating Quality 4.0 with AI and ML.

This manuscript contributes to the existing body of knowledge by introducing a novel framework that integrates Quality 4.0 principles with organizational strategic planning processes, laying the groundwork for Strategic Planning 4.0. These findings have the potential to guide future research, influence industry practices, and foster innovation and discovery, enabling organizations to thrive in the dynamic Industry 4.0 landscape.

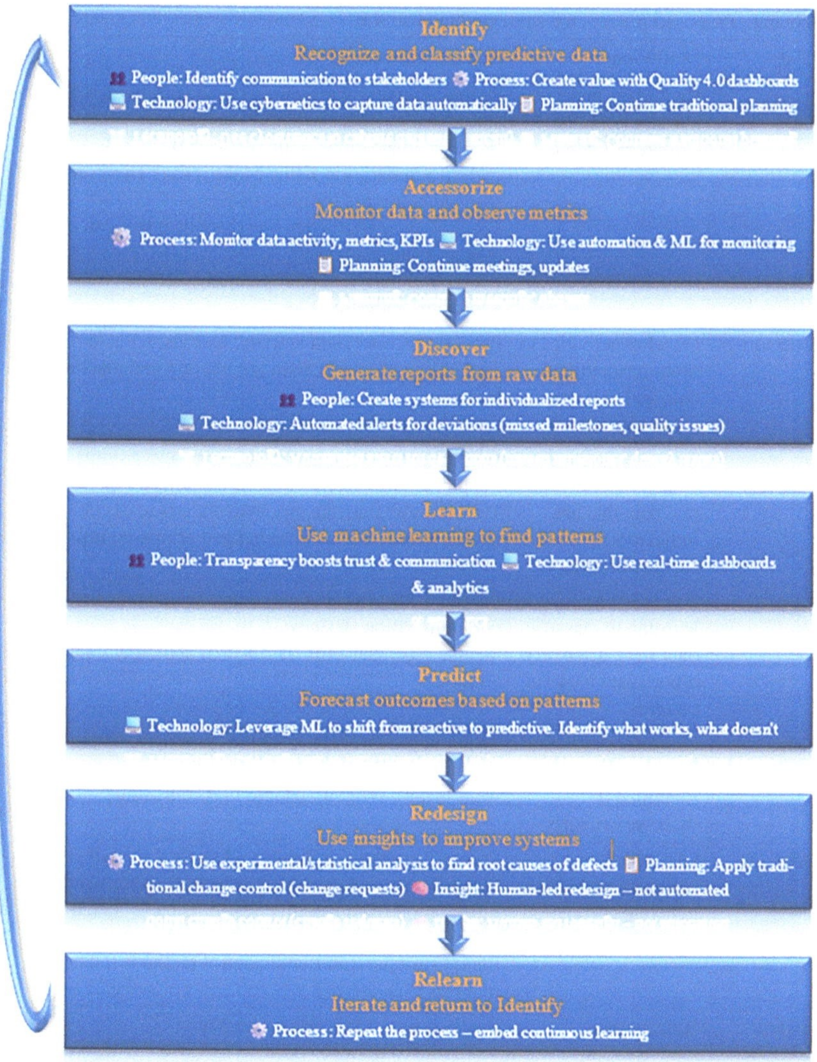

Fig. 6.1 Visual framework of the proposed process for strategic planning 4.0

6.5 Implications for Theory and Practice

6.5.1 Limitations of Research

While this manuscript contributes to research as it relates to quality management and strategic planning, it encountered limitations and areas that merit further investigation. First, this manuscript focuses exclusively on the integration of Quality 4.0

Table 6.2 Proposed process for strategic planning 4.0

Stage	Definition	Activity examples
Identify	Recognize and classify data that can predict defects	People: identify communication to stakeholders that can be automated (such as reminders and alerts) Process: create value with quality 4.0—use dashboards with live updates on KPIs. Make it fun! This is not just for management; it is for everyone! Technology: cybernetics—is there data that can be captured automatically? Plan: continue traditional planning methods (resource allocation, milestones, etc.)
Accessorize	Develop activities to monitor identified data and observes those activities	Process: monitor data activity, metrics, KPIs, etc. Technology: monitor with automation and machine learning Plan: continue traditional management methods (meetings, updates)
Discover	Create reports from raw data that are indicators of success or failure	People: create systems for individualized reports Technology: generate automated alerts to deviations from quality standards such as changed or missed milestones
Learn	Use machine learning to classify data from discovery reports and look for patterns	People: be transparent about data to enhance trust and communication strategy Technology: real-time dashboards and other data analytics are key
Predict	Optimize prediction based on patterns identified in discovery reports	Technology: leverage machine learning to be predictive, not reactive. What is resonating? What is working and what is not? What do people find confusing?
Redesign	Use relevant observational data to generate correlation	Process: experimental data and statistical analysis are used to determine root causes of defects. In other words, use information from machine learning to evaluate quality and effectiveness. This redesign stage is not automated Plan: Continue traditional control methods (change requests)
Relearn	Relearn data	Process: repeat this entire process

principles into the strategic planning process. Notably, it did not discuss in detail the broad value of strategic planning, the specific quality of an organization's strategic plan, or the intricacies of the implementation. Instead, its primary objective is to optimize the strategic planning process by leveraging the transformative capabilities of Quality 4.0. Second, the framework design targets mid-size and large organizations with the necessary resources and complexities that demand a predictive and automated approach. Small operations with limited funds and resources may find it impractical. Finally, Escobar [12] recognized that the 'evolution of problem-solving' framework remains in its early stages, while the field of practical applications for

Quality 4.0 continues to expand [3]. Further research is needed to refine a framework for Quality 4.0 and its application to all phases of strategic planning.

6.5.2 Key Takeaways for Management

In the era of Industry 4.0, defined by digital technology, machine learning, and artificial intelligence, organizations must evaluate their readiness for Quality 4.0 adoption to thrive [7]. It is imperative for organizations to find ways to reduce manual data collection and leverage automation and machine learning [16]. While Escobar [12] noted challenges in developing and adopting a concrete solution for monitoring and control that leverages Quality 4.0, modern quality management should be integrated in strategic planning processes to drive innovation and success in today's technological landscape [5, 9].

George [1] introduced the people-process-plan framework serves as an additional tool in strategic planning, noting that it cannot replace the need for formality, comprehensiveness, and effective stakeholder management in the planning process. Chiarini and Kumar [18] offered a simple Quality 4.0 framework, people-process-technology. The below key takeaways are thus organized as people-process-technology-plan for Strategic Planning 4.0.

- **People**: Stakeholders are key in identifying and categorizing predictive defect data. To enhance strategic planning, create a roadmap that seamlessly integrates Quality 4.0 principles, ensuring alignment with organizational goals.
- **Process**: Develop automated activities to monitor data, metrics, and KPIs, and create reports for interpretation. The systematic flow starts with identifying potential defects and emphasizes continuous learning. While embracing Quality 4.0 approaches, continue to incorporate traditional elements (e.g., timelines, resource allocation, and specific milestones).
- **Technology**: Allocate resources for technology infrastructure (data analytics, automation, machine learning) and upskill workforce in data analysis and technology proficiency. Leverage automation and machine learning for deviations from quality standards and missed milestones, real-time monitoring, up-to-date dashboards, classifying data for discovery reports., and creating algorithms enhance efficiency and accuracy.
- **Plan**: Continue traditional approaches to strategic planning but cultivate an environment that values people and automated processes.

In summary, organizational management should prioritize leveraging automation and machine learning to streamline manual data collection and other monitoring and control activities as part of their strategic planning process to thrive in the modern business landscape.

6.6 Conclusions

The next phase of quality management, Quality 4.0, leverages traditional quality management approaches with automation and machine learning. This manuscript discussed project quality management, Quality 4.0, and strategic planning, and offered a possible framework that integrates modern technology with organizational strategic planning processes, or Strategic Planning 4.0. This manuscript highlights the benefits of integrating modern technology with strategic planning processes, though limitations and challenges remain.

Quality 4.0 is a developing area of research and a definitive framework similar to traditional quality management approaches such as TQM has yet to emerge. This manuscript recommends a framework primarily geared for medium and large organizations with the budget and resources to invest in new technology. This manuscript focuses solely on integrating Quality 4.0 in the strategic planning process and does not delve into evaluating the quality of specific strategic plans or providing best practices for implementation.

To comprehensively grasp the interplay between Quality 4.0 principles, the people-process-technology nexus, and their impact on business outcomes and customer satisfaction, future research should delve into both qualitative and quantitative investigations. Another valuable avenue for further research lies in investigating the integration of Quality 4.0 within strategic planning and its implementation.

Strategic planning remains the bedrock of organizational strategies by cultivating a unified vision, reinforcing responsibility and accountability, and ultimately driving successful outcomes. In today's dynamic enterprise environment, aligning the strategic planning process with best practices is paramount. Through the deliberate integration of Quality 4.0 principles, forward-thinking managers can cultivate an organizational culture that values people, optimizes processes, and embraces technological advancements. This strategic alignment not only propels the organization toward "Strategic Planning 4.0," but also positions it at the forefront of innovation and sustainable growth.

References

1. George, B.: Successful strategic plan implementation in public organizations: connecting people, process, and plan (3Ps). Public Adm. Rev. **81**(4), 793–798 (2021). https://doi.org/10.1111/puar.13187
2. Cote, C.: Why is strategic planning important? HBS Online, Business Insights Blog (2020). https://online.hbs.edu/blog/post/why-is-strategic-planning-important
3. De Oliveira, D.M., Alvelos, H.R., Rosa, M.J.: Quality 4.0: results from a systematic literature review. TQM J. (2024). https://doi.org/10.1108/tqm-01-2023-0018
4. Edwards, R.L., Wollner, S.B., Weddle, J., Zembrodt, J.W., Birdwhistell, M.D.: Diagnosing and resolving conflict created by strategic plans: where outreach strategies and execution meet at an academic health center. Hosp. Top. **95**(3), 72–78 (2017). https://doi.org/10.1080/00185868.2017.1301172

5. El Zein Badawi, S.M.: Transformation of higher institute of education by integrating quality assurance and strategic planning: the experience of Ahfad University for Women—Sudan. Ahfad J. **37**(2) (2020)
6. Albon, S.P.: Strategic planning in an educational development centre: Motivation, management, and messiness. Coll. Essays Learn. Teach. **9**, 207–226 (2016). https://doi.org/10.22329/celt.v9i0.4427
7. Alzahrani, B., Bahaitham, H., Andejany, M., Elshennawy, A.K.: How ready is higher education for quality 4.0 transformation according to the LNS research framework? Sustainability **13**(9), 5169 (2021). https://doi.org/10.3390/su13095169
8. Krivokuca, M.: The concept of quality 4.0 continues to evolve. Qual. Mag. (2022). https://www.qualitymag.com/articles/97258-the-concept-of-quality-40-continues-to-evolve
9. Soltani, M.D.: The impact of total quality management on firm competitiveness by the mediating role of product and process innovation. J. Bus. Adm. **14**(28), 1–30 (2022)
10. Project Management Institute: A guide to the Project Management Body of Knowledge (PMBOK Guide) – Sixth Edition/Agile Practice Guide Bundle. Project Management Institute, Newtown Square (2017)
11. Lotich, P.: 12 steps to implementing a total quality management system. The Thriving Small Business (2023). https://thethrivingsmallbusiness.com/implementing-a-quality-management-system
12. Escobar, C.A., McGovern, M.E., Morales-Menendez, R.: Quality 4.0: a review of big data challenges in manufacturing. J. Intell. Manuf. **32**(8), 2319–2334 (2021). https://doi.org/10.1007/s10845-021-01765-4
13. Haigh, T.: Artificial intelligence then and now: from engines of logic to engines of bullshit? Commun. ACM (2025). https://cacm.acm.org/opinion/artificial-intelligence-then-and-now/
14. Liu, H., Liu, R., Gu, X., Yang, M.: From total quality management to quality 4.0: a systematic literature review and future research agenda. Front. Eng. Manag. **10**(2), 191–205 (2023). https://doi.org/10.1007/s42524-022-0243-z
15. Maganga, D.P., Taifa, I.W.R.: Quality 4.0 conceptualization: an emerging quality management concept for manufacturing industries. TQM J. **35**(2), 389–413 (2023). https://doi.org/10.1007/s42524-022-0243-z
16. Emblemsvåg, J.: On quality 4.0 in project-based industries. TQM J. **32**(4), 725–739 (2020). https://doi.org/10.1007/s42524-022-0243-z
17. Project Management Institute: A guide to the Project Management Body of Knowledge (PMBOK Guide) – Seventh Edition. Project Management Institute, Newtown Square (2021)
18. Chiarini, A., Kumar, M.: What is quality 4.0? An exploratory sequential mixed methods study of Italian manufacturing companies. Int. J. Prod. Res. **60**(16), 4890–4910 (2022). https://doi.org/10.1080/00207543.2021.1942285
19. Fernández-Loría, C., Provost, F., Han, X.: Explaining data-driven decisions made by AI systems: the counterfactual approach. Manag. Inf. Syst. Q. **46**(3), 1635–1660 (2022). https://doi.org/10.48550/arXiv.2001.07417

MIX
Papier aus verantwortungsvollen Quellen
Paper from responsible sources
FSC® C105338

If you have any concerns about our products,
you can contact us on
ProductSafety@springernature.com

In case Publisher is established outside the EU,
the EU authorized representative is:
Springer Nature Customer Service Center GmbH
Europaplatz 3, 69115 Heidelberg, Germany

Printed by Libri Plureos GmbH
in Hamburg, Germany